THE SCIENCE OF BATTLESTAR GALACTICA™

THE SCIENCE OF BATTLESTAR GALACTICA

PATRICK DI JUSTO
AND
KEVIN R. GRAZIER

John Wiley & Sons, Inc.

Published by John Wiley & Sons, Inc., Hoboken, New Jersey
Published simultaneously in Canada

Design by Forty-Five Degree Design LLC

Credits appear on page 293.

Library of Congress Cataloging-in-Publication Data:
Di Justo, Patrick, date.
The science of Battlestar Galactica / Patrick Di Justo, Kevin Grazier.
p. cm.
Includes bibliographical references and index.
ISBN 978-0-470-39909-5 (pbk.); ISBN 978-0-470-88202-3 (ebk);
ISBN 978-0-470-88203-0 (ebk); ISBN 978-0-470-88204-7 (ebk)
1. Technology–Popular works. 2. Science–Popular works. 3. Astronomy–Popular works. 4. Science fiction, American. 5. Battlestar Galactica (Television program : 1978–1979) 6. Battlestar Galactica (Television program : 2003) 7. Battlestar Galactica (Television program : 2004–2009) I. Grazier, Kevin Robert, date. II. Title.
T47.D54 2010
600–dc22

2010018485

Printed in the United States of America

10 9 8 7 6 5 4 3 2 1

For my mother and the memory of my father
—P. D.

To everybody who made
Battlestar Galactica a reality
—K. G.

CONTENTS

PART TWO
The Physics of *Battlestar Galactica*

PART THREE
The Twelve Colonies and the Rest of Space

PART FOUR
Battlestar Tech

FOREWORD

"I'm going to tell the best possible story, and I don't care how many geeks I have to offend to get it done."

This sentiment, in various incarnations, is frequently expressed by television writers. What they mean is that they're willing to ignore science, the what-would-really-happen, in order to tell the most compelling story. And, honestly, that makes sense. A TV writer's job is to entertain, after all. It's *30 Rock*, not *Igneous Rocks*. And it is certainly true that the tension in your real-time nail-biter of a detective drama might dissipate somewhat during that realistically long wait while the lab patiently sequences DNA. Settling on a less-than-dramatic story just so that the tiny percentage of viewers who know the science aren't tempted to throw things—it just doesn't seem like a winning strategy.

But the two priorities do not have to lie in opposition to each other. At *Battlestar Galactica* (also referred to in this book as *Battlestar, Galactica*, and *BSG*), writers were not allowed to jettison science for the sake of story. Other than in specific instances of intentionally inexplicable phenomena, science was respected. In this way, *Battlestar Galactica* resembled the classic tradition of science fiction novels, which didn't see science as a hindrance to the story but as a springboard.

The first episode of *Battlestar* that I cowrote was called "The Passage," and it was all about a risky mission in which pilots escort an instrument-blind fleet through a high-radiation space phenomenon called a "globular cluster." The mission was complex in terms of tactics as well as science, a tough one-two punch. My initial instinct was to simplify the mission, the dangers, the science—to simplify something—because I simply couldn't believe that the fragile tension of a forty-two-minute story could bear for seven of those minutes to be exposition. I wasn't convinced that it would work, in fact, until I saw the completed episode. The meticulous set-up of the strategy and dangers rings true because it is true. At least, it is true to the extent that I captured what I was being educated about during the writing process. (Did you know there's a difference between radiation and radioactivity? Oh, well, that's just me that didn't know, then.)

The sentiment of story over science with which I started this piece still echoed in my ears, even after "The Passage," and there was a period during which I had to learn the lesson repeatedly. When I needed something to malfunction on a Raptor, I was surprised to find that I actually had to investigate which of the Raptor's functions were controlled by which mechanisms and where they were located. When I needed to describe the effects of death in a vacuum, I needed to resist rumor ("Your lungs pop out of your nose!") and find out what really happens. (Your lungs don't pop out of your nose or anywhere else interesting.) And don't get me started on "single-event upsets."

Sometimes we writers ignored the advice we were given—either because there is a limit on the stretchiness of a teleplay to accommodate explanation, or because we really were dealing with a situation that intentionally defied explanation, or just because of limits on our time and skill. But we really tried, and the truism of story versus science seemed less true every time I tested it. In fact, it starts to seem absurd. Truth, generally defined as "emotional truth," is a huge part of what makes a good story. Is it so surprising that other kinds of truth would also support, not undermine, the same goal? As the child of a chemist, I'm startled it took me so long to realize it.

The man supplying the *Battlestar* writers with our technical and scientific information was Kevin Grazier. In the time that I was writing for the show he personally advised me on: algae, globular clusters, fuel processing, Raptor thrust and steering mechanisms, CO_2

scrubbers, the effects of a vacuum on a human body, SEUs, hypoxia, hypothermia, constellations, the distances of space, Jumping, and a lot of things I've already forgotten. If the Cylon base ship toilet in one of my first drafts had survived to a production draft, I suspect we would've had a talk about it, too. (I saw it as a sort of living pulsing suction-cup-shaped protuberance.)

The point of *Battlestar Galactica* was not, ultimately, science. It was a show about the human condition, hope, and moral grayness. But science didn't distract us from telling the stories we wanted to tell. It helped us. And Kevin helped us get the science right. So enjoy this volume—if Kevin says it's true, it must be so.

So Say We All!

Jane Espenson
Co-executive Producer
Los Angeles

ACKNOWLEDGMENTS

Kevin says: Common wisdom holds that everybody who works in Hollywood does so because they caught their "big break." For mine I would like to thank Bryan Fuller for "pitching" me as science advisor to *Battlestar Galactica* executive producer Ronald D. Moore. On a related note, I can't thank Ron enough for taking a leap of faith and giving this longtime sci-fi nerd the opportunity and the honor to work on the best show on television, especially after what was likely the shortest job interview in recorded history. I'd also like to thank Patrick Di Justo, Giles Anderson, and Connie Santisteban for including me in this project in midstream. Thank you to executive producer David Eick and the cast and crew—in particular the writing staff—of *Battlestar Galactica*. Not only did their efforts make this book possible, but everyone was a joy to work with over the span of four they-went-far-too-quickly seasons. Finally, I'd like to thank Kendra "Not Shaw" Penny for all her sacrifices that made my contribution to this book possible. So Say We All!

Patrick says: I'd like to thank our Wiley (and wily) editor, Connie Santisteban, who safely shepherded this project through the times when we were under Cylon attack; my agent, Giles Anderson, who convinced me to write this proposal as an exercise and then turned it into a book; the folks at *Wired* magazine, including but not limited to Chris Baker, Rob Capps, Erik Malinowski, Joanna Pearlstein, Adam Rogers, and Nick Thompson; everyone in the Facebook group "The Science of *Battlestar Galactica*"; the BSG–Park Slope triumvirate: Erica Blitz (aka "ProgGrrl") of the fansite Galactica Sitrep, superfan and

Daily Show writer Rob Kutner, and BSG guest star John Hodgman; my fellow Brooklyn BSG panelists: John Brooks, Ajay Singh Chaudhary, Shane Froebel, and Laura Lee Gulledge; book guru Ellie Lang; Megan Kingery of Science House; pop culture writer Racheline Maltese; the wonderfully supportive people of The WELL's <byline> and <sftv> conferences; and my siblings Andy Di Justo and Melissa Perdock.

This book could not have been completed without Emily Gertz. During the writing of this book, Emily made sure I ate, made sure I slept, stuck with me in the emergency room when my zeal to produce this book put me there, and on occasion would remind me to repeat to myself it's just a show—I should really just relax.

INTRODUCTION

Moore's Law

It has been called the best show on television, and as real as science fiction gets. It has dealt with issues of religious freedom, basic human rights, patriotism, terrorism, genetic engineering, and the ultimate science fiction question: What does it mean to be human?

The show is *Battlestar Galactica*, a twenty-first-century reimagining of the classic 1970s television series with the same basic storyline: humanity inhabits twelve different planets, known as the Twelve Colonies. The ancient birthplace of their species, a planet called Kobol, is a distant memory. Some fifty years prior to the beginning of the series, Colonial scientists created a race of robots they called Cylons, who eventually attained sentience and rebelled against their creators. After a long and bloody war, which ended in a stalemate/cease-fire, the Cylons withdrew, presumably to find a planet to call their own. Eventually the robots evolved from purely mechanical creatures into biologically engineered creatures nearly indistinguishable from the Colonials themselves.

At the beginning of the miniseries, these humanoid Cylons insinuate themselves stealthily into Colonial culture, circumvent the defenses of the Twelve Colonies, and wipe out almost all of humanity. The only humans who are spared are a handful on the various planets who are able to escape and those fortunate enough to be on spacecraft during the attack: roughly fifty thousand humans on sixty vessels. This Rag Tag Fleet has only the aging battlestar *Galactica* for protection. Now in need of a home, they decide to search for the semimythical planet Earth, with the Cylons in relentless pursuit.

The new *Battlestar Galactica* (BSG to its fans) eschews the meaningless whiz-bang shoot-'em-ups of the earlier show,* choosing to focus instead on character development and conflict, both internal and external. Sometimes the characters are so developed that it becomes difficult to tell exactly who the good guys are. The executive producer David Eick has said that "We're not doing our jobs if, at least once a week, the viewer doesn't ask, 'Am I rooting for the wrong team?'"

A case in point comes in the third-season episode "Precipice." Colonel Saul Tigh, a career military officer in the Colonial Fleet, is trapped on New Caprica, a barely habitable planet the Colonists are calling home. The Cylons have occupied the planet, and Tigh has become a full-fledged resistance fighter/terrorist against the occupation government. When another character questions Tigh's use of terror against civilians, Tigh tells him exactly where things stand: "Which side are we on? We're on the side of the demons, Chief. We are evil men in the gardens of paradise, sent by the forces of death to spread devastation and destruction wherever we go. I'm surprised you didn't know that." We, the audience, found ourselves supporting Tigh, silently advocating the same tactics we condemned when the Cylons used them only a few episodes before.†

Many of the elements of the show have parallels to real life: Caprica, the most advanced planet of the Twelve Colonies, represents America.‡

---

*For example, each episode does *not* automatically have a space battle.

†Of course, the genius mind-frak of that scene is that we were manipulated into agreeing with one crypto-Cylon talking to two other crypto-Cylons about fighting the regular Cylons, only nobody knew that yet.

‡But does it?

Lee Adama, Admiral William Adama, and Kara "Starbuck" Thrace.

Cylon models Eight and Six with scientist Gaius Baltar (middle).

The Cylons are fundamentalist religious terrorists.* The Cylon attack on the Twelve Colonies invokes memories of 9/11.† The swearing-in of President Roslin is filmed in the same manner as that of President Johnson following JFK's assassination in Dallas.‡ The list goes on and on. We understand that this same story of the clash of civilizations could be told—and is being told every day on the nightly news—without the trappings of space flight and robots. For that reason, *Battlestar Galactica* has also been called a science fiction show without science.

Ron Moore is on the record as saying that when he worked on the various *Star Trek* TV series, the intense level of Roddenberrian technolove for the Starship *Enterprise* actually hindered story development when the immense level of background information established that the *Enterprise* worked in a certain way that contradicted the story they wanted to tell. As Moore famously complained in the January 2000 issue of the magazine *Cinescape* about *Star Trek Voyager*:

> In the premise this ship was going to have problems. It wasn't going to have unlimited sources of energy. It wasn't going to have all the doodads of the *Enterprise*. It was going to be rougher, fending for themselves more, having to trade to get supplies that they want. That didn't happen. It doesn't happen at all, and it's a lie to the audience. I think the audience intuitively knows when something is true and something is not true. *Voyager* is not true. If it were true, the ship would not look spic-and-span every week, after all these battles it goes through. How many times has the bridge been destroyed? How many shuttlecrafts have vanished, and another one just comes out of the oven? That kind of bullshitting the audience I think takes its toll. At some point the audience stops taking it seriously, because they know that this is not really the way this would happen. These people wouldn't act like this.

---

*But are they?

† Well, yes and no.

‡ Okay, that one is real.

In that quote you can practically see the rough beast of a resurrected *Galactica* slouching toward Ron Moore's forebrain to be born.

Keep in mind that *Galactica* is just barely science fiction, at least as too many teenagers and television executives define the term. There are no alien civilizations, no dwarfs, no elves, no wizards, no sentient trees, no magic rings, no magic wands, no magic spells, no magic cloaks, no cloaking devices, no lasers, no phasers, no photon torpedoes, no time travel. Instead, there is blood and sweat and spit and heartache and anger, guns that shoot lead instead of light, and the pervasive fear each character carries that their slightest mistake could literally mean the end of their species.

Galactica is science fiction in the greatest sense of the term, in that *Galactica* is a show about *ideas*. It is based, in part, around the ultimate science fiction question: What does it mean to be human? By being science fiction, *Galactica* allows its producers and writers to discuss issues that would rarely be allowed on American commercial television. What is each person's relation to God? What is the individual's responsibility to the government, and vice versa? When, if ever, is it acceptable to torture? When, if ever, is it acceptable to curtail basic human rights? Yet while it is true that *BSG* is not a technophile's dream, the science that *does* exist in the show serves to illuminate the use of science and technology in our own lives. The debate over whether to use Cylon blood to cure human cancer is essentially an argument about the pros and cons of stem cell therapy; the development of "The Farm" by the Cylons lays out the pro and con arguments for abortion, choice, and parental responsibility. Like all good science fiction, *Battlestar Galactica* takes us millions of miles away from Earth for the sole purpose of letting us turn around to see ourselves from a different perspective.

The Science of Battlestar Galactica is affected by what might be called Moore's Law. Taken from his thoughts as presented in his *Cinescape* interview, and reiterated in his blog entry of February 1, 2006, the law can be stated thusly: "We *always* tried to make drama work with science on BSG, but when push comes to shove, drama wins."

Moore popularized the word "technobabble"—the meaningless sciencey-sounding words a writer puts into an SF script to "explain" a scientific plot problem and to keep the geeks happy—when he was working on *Star Trek: The Next Generation*. To keep the technobabble mafia from muscling in on *Galactica*, Moore simply decided not to explain how anything worked. It allowed him the freedom to invent Cylons who are indistinguishable from humans, at least to the level of standard medical tests, yet who can also control a Colonial computer by mainlining a fiber optic cable. Great for storytelling. Drives us geeks crazy if not explained.

For that reason, we've found it necessary to establish the three laws of *The Science of Battlestar Galactica*.

The First Law of *The Science of Battlestar Galactica* takes care of the lack of scientific explanation, the unresolved plot issues, and the Fleet's endless supply of whiskey: "If you're wondering how they eat and breathe, and other science facts, just repeat to yourself, 'It's just a show, I should really just relax.'"[1]

The Second Law of *The Science of Battlestar Galactica* is a quote from Carl Sagan: "Space is mostly empty. That's why it's called 'space.'"

The Third Law of *The Science of Battlestar Galactica* is: "All this has happened before and will happen again." Don't lose sight of this.

Through the course of this book, we will assume you've seen all eighty-four hours of BSG, so don't look for warnings: all plotlines will be spoiled, all secrets will be examined. Our own technobabble will be limited to a few terms. "Earth," in these pages, *always* refers to the planet on which this book was written. "Dead Earth" *always* refers to the radioactive planet the Colonial Fleet found at the end of season four. "Earth II" *always* refers to the preindustrialized planet the Fleet found in the finale. And the Cylon attack on the Twelve Colonies serves as the dividing line between eras in the BSG universe: we'll be calling them BF (Before the Fall) and AF (After the Fall). For example, the Cylons rebelled against the Twelve Colonies about 50 BF, the Fleet found Earth about 3AF, and so on.

You've bought the ticket; prepare to enjoy the ride!

PART ONE

LIFE HERE BEGAN OUT THERE

CHAPTER 1

Are You Alive?

The original 1978 *Battlestar Galactica* began with the narrated words "There are those who believe that life began out there." In the 2003 miniseries, Commander William Adama cites the opening lines of the sacred scrolls of the Colonial civilization, which reiterate the idea. Here on Earth, many religions and more than a few scientists also suggest that life was brought to a lifeless Earth from elsewhere in the universe.

What is life? How did it begin? What is the story of the origin of life?

Life has always been a tricky concept to define. Many people and cultures have tried as soon as they were able to ask the question.

Number Six: Are you alive?
Military Liaison: Yes.
Number Six: Prove it.
—Miniseries, Part I

Because life seemed to be something magical, something that could not be explained by ordinary means, many of them looked to spiritual or extra- or paranormal ways to explain life. Life, for many

cultures, had a certain divine spark to it, a spark that could not be understood by ordinary humans.

Gradually, as reason came to replace mysticism, life developed a different definition. The mystical aspects of life were lost and scientists viewed life as a continuing process—or rather, several processes: ingestion, excretion, growth, reproduction, response to stimuli, and death.

The idea that "life is what it does" was incomplete; a skilled debater could always find exceptions. What if, for instance, a living creature existed on a much slower (or much faster) time scale than ours? Pick up a rock. Look at its various layers, or embedded crystals. Now ask yourself why you *know* that this rock is not alive. The rock in your hand might very well be ingesting, excreting, growing, reproducing, reacting, and dying—but on a scale of millions of years. At that rate, you'd never detect any of its activity. The rock would seem to be dead.

Or take the case of mayflies. They live their entire lives from birth to death in one day, just long enough for us to understand what they're up to. But would we be able to recognize creatures that lived their entire lives in one hour? One minute? One second? A fraction of a second? To us, their entire existence would appear like a flash of faraway lightning, so brief we wouldn't be sure it was there at all. To them, we would be like rocks, unmoving and unchangeable, the stable backdrop against which they go about their lives.*

Since we're dealing with a science fiction show that takes place in space, we should perhaps use a definition of life developed by people looking for life in outer space. In the early 1990s, a NASA astrobiology panel defined "life" as *a self-sustaining chemical system capable of undergoing Darwinian evolution*.† It's not perfect, but it'll do.

The best scientific explanation available to the Colonials is that life would have begun on Kobol sometime around 3.8 billion years ago, when the planet had finally solidified, and water vapor in the atmosphere precipitated out to form liquid water on the surface. The water acted as a solvent for other chemicals, and the constant intermixing

* This idea was explored very well in the *Star Trek* TOS episode "Wink of an Eye."

† That innocuous definition is under debate to this day, even within NASA itself.

Gaius Baltar and Six hold human/Cylon hybrid Hera, in a vision.

Caprica Six and Baltar on Caprica, before the attack.

created a bouillabaisse of various chemicals and molecules just waiting to be put into the right order.

According to our current understanding, somehow—we still don't know the details—the raw chemicals of life (usually nothing more elaborate than hydrogen, carbon, oxygen, and nitrogen) came together in various configurations in millions of places in the oceans over millions of years. In all this turmoil, eventually one configuration became able to produce crude copies of itself. This particular molecule, by the virtue of copying itself, eventually came to dominate the early environment. By continually mutating and adapting, some copies became more and more efficient at copying themselves, and others became less efficient. The ones that were less efficient at copying themselves were by definition unfit, and they quickly died out. The ones that survived also developed ways to find sources of energy in the immediate vicinity. By acquiring energy from the environment, the process became self-sustaining, and met NASA's definition for life.

As Kobol formed from smaller asteroids and comets, the last pieces of comet ice to arrive became the water and gases that made Kobol's oceans and atmosphere. These comets were essentially small chemistry sets. In addition to water, comets are reliably thought to have seeded the newborn planet with methane (CH_4), ammonia (NH_3), hydrogen sulfide (H_2S), phosphate (PO_4^{3-}) radicals, carbon dioxide (CO_2) and carbon monoxide (CO). With those chemicals you can build a kingdom of life.

It is very likely that Kobol's first atmosphere was primarily made of five gases: nitrogen, carbon dioxide, carbon monoxide, ammonia vapor, and methane. There was no oxygen in this atmosphere, hence no ozone layer.* With no ozone layer there would be almost no filter for ultraviolet radiation from the sun, so almost all of it would reach the surface. To anyone who is already alive, this would be a bad thing, but on a sterile Kobol before the creation of life, it could be just what was needed.

The discovery of lightning in the clouds of Venus and Jupiter (whose atmosphere is quite similar to that of early Kobol) means

*The chemical formula for ozone is O_3.

that it is very likely that Kobol's early atmosphere was also rent by the occasional lightning bolt. Both ultraviolet radiation and lightning are mechanisms that could have put enormous amounts of energy into Kobol's oceans and atmosphere. Chemical reactions need some sort of energy to get going, and Kobol had that in abundance. When you have millions of years to work with, nearly any type of energy is enough to make some form of organic compound out of those ingredients.

This isn't just speculation. In 1953, two scientists at the University of Chicago, Stanley Miller and Harold Urey, created complex organic molecules out of simple elements. Miller and Urey built a small planetary ecosystem in their laboratory. They started with a flask filled with liquid water and the appropriate salts and chemicals (CH_4, NH_3, H_2) dissolved in it to represent Earth's early ocean. They connected that flask to a series of glass tubes and reservoirs that contained the gaseous mixture that represented Earth's early atmosphere. In the glass reservoir that held the "atmosphere," they placed two electrodes, positioned just so to create a spark gap. A condenser at the bottom of the "atmosphere" precipitated the water vapor and closed the loop by sending it back into the "ocean." Miller and Urey gently warmed their ocean to evaporate some of the water into the atmosphere, and then ran a simple electric spark through the resulting gas to simulate the action of lightning.

They ran the experiment for a week and watched as the water in their ocean turned pink and then brown. After a week they analyzed the ocean fluid and discovered that it was full of complex organic molecules, the most exciting of which were the simple amino acids glycine and alanine, two of the building blocks of protein. Amino acids[1] are not life per se, but without amino acids there would be no proteins or enzymes, and without proteins or enzymes there would be no life as we know it. Subsequent similar experiments by Miller and Urey and by other scientists had even more interesting results. By adding ammonium cyanide to the mixture, it is possible to synthesize adenine, a nucleic acid and one of the building blocks of DNA! Think about that—one of the pieces of the control mechanism that makes you the unique being that you are can be made on practically any planet that has a similar atmosphere, similar ocean, similar energy sources.

Even further experiments along the Miller-Urey line discovered that with a slightly different concentration of starter materials, one could create fatty acids or lipids. Because of the nature of fats, they naturally tend to collect themselves into spherical globules, which is nearly perfect for encapsulating and concentrating various types of material—such as the material that makes up a living cell. Every living cell is essentially a bubble of fat with sweetened protein-water gunk inside it, and every piece of that cell could have come from a few simple chemicals and a spark.

CHAPTER 2

The Cylons: Man or Machine?

According to Daniel Graystone, CEO of Graystone Industries and inventor of the first Colonial Cylon, "Cylon" is a backronym for Cybernetic Lifeform Node.* Cylons come in at least three distinct versions—so distinct we'd do well to give each group a discrete name so that there's no confusion as to what we mean when we talk about them.

Humanoid Cylons

Humanoid Cylons: There were thirteen different models in this class, divided into two groups: the Final Five (who should really be called the First Five) and the Significant ~~Seven~~ ~~Eight~~ Seven. For the rest of this book we'll be referring

*Thus capturing the "Worst SF Backronym" title, previously held by C.H.U.D.

Gaius Baltar with Cylons Tory Foster and Six.

Final Five Cylons Ellen and Saul Tigh.

Eight and Six with four of the Final Five: Galen Tyrol, Tory Foster, Saul Tigh, and Sam Anders.

to the known twelve humanoid models as just Cylons,* unless we need to specify one or the other.

Cylons are sentient synthetic biological creatures, possessing all the attributes of life except evolution by natural selection. They have been designed to be so biochemically similar to the people of the Twelve Colonies that they can pass pretty stringent standardized medical tests without being detected. They have—or they have been programmed to behave as if they have—all the emotions, intelligence, and weaknesses of the people they were based on (something that pisses off the Cylon known as Brother John Cavil to no end).

The Final Five Cylons, the ones we know as Tory Foster, Ellen Tigh, Colonel Saul Tigh, Chief Petty Officer Galen Tyrol, and Ensign Samuel

*In the show they're also called Skin Jobs. "Skin job" is a direct homage to the film *Blade Runner*, which explored similar issues of engineered sentient creatures and the difference between what is human and what is not. In that movie it is an extremely offensive term.

Anders, are the only survivors of the first generation of humanoid Cylons created by the inhabitants of the planet Kobol more than four thousand years before the destruction of the Twelve Colonies. These humanoid Cylons were able to breed, and eventually formed themselves into a thirteenth tribe called, appropriately enough, Cylon. This tribe of Cylons was on moderately good terms with their creators, but at some point in their history they found it advantageous to leave Kobol and seek their own planet, which they called Earth. They were destroyed two thousand years later in an all-out nuclear attack by the sentient Centurion-type robots they themselves had created. The Final Five, forewarned of the upcoming cataclysm by the same "angels" who later appeared to Dr. Gaius Baltar, Caprica Six, Captain Kara "Starbuck" Thrace, and perhaps others,* managed to survive the attack and used a relativistic spacecraft to find their way to the Twelve Colonies, though this revelation comes from a rambling Sam Anders with a bullet in his brain, and could be considered of questionable veracity.

There, they found a situation similar to the one that had befallen their home planet: mechanical Cylon Centurions in the Twelve Colonies had turned on their creators and started a war against them. More importantly, these Centurions had begun to experiment with creating their *own* synthetic biological creatures. According to Ellen Tigh, she and the other Four agreed to help the Centurions to make humanoid Cylons if they would end the war against the Colonies. The result was the group of Cylons we call the Significant Seven.

All humanoid Cylons look, feel, sound, smell, and taste the same way we do, as evidenced by the number of Colonials who have had sex with Cylons† without even realizing it. At least one model of Cylon,

---

*Sam Anders tells us this just before he's wheeled into surgery to remove a bullet from his head in the season four episode "No Exit": "Back on Earth the warning looked different to each of us. I saw a woman; Tory saw a man."

†Gaius Baltar is at the top of the Colonial Pimp Daddy list for gettin' a little sump'n sump'n with at least three Cylon models: various Sixes, D'Anna Biers (a Three), and Tory Foster. The others are:

- Karl Agathon (Sharon and Boomer, both Eights)
- Specialist Calley (Chief Tyrol)
- Starbuck (Anders)
- Admiral Cain (Gina Inviere, a Six)

the one we know as Saul Tigh, can age. And probably most important for the overall arc of the show, Cylons can breed with Colonials.

And oh, how the Cylons are obsessed with sex! Sex infuses almost all of their interpersonal relationships, and Cylons spend their days as horny as a roomful of science fiction fans at a *Galactica* convention. The Cylons (at least the female ones) display some form of luminescence along their spine at the moment of orgasm, which must have made Baltar's threesomes with D'Anna and Six look like a laser light show.

Significant Seven Cylons seem to have problems reproducing biologically with each other. Number Six explained to Baltar their belief that this apparent infertility is due to the inability to feel real human love between the partners. We have evidence that at least one model of Cylon, Number Eight, can be fertilized by Colonial males, and another model, Number Six, can be fertilized by Saul Tigh, a member of the Final Five. Tigh was using his own Cylon projection system to imagine that he was making love with his beloved wife Ellen instead of Six, so maybe there *was* real love in the act.

Not to be too cynical, but with most species lack of love is usually not enough to prevent conception. Why do Cylons have such trouble impregnating other Cylons? It may be that the Cylon gene pool is so small, and Cylons are so genetically alike, that fetuses spontaneously abort. Maybe Cylon eggs and sperm are too fragile to survive with each other. Think of Cylon gametes as if they were half-dead batteries, and Colonial gametes as fresh batteries. If you power a gadget with one half-dead battery and one fresh battery, the gadget will most likely work. If you use two half-dead batteries, the gadget will most likely not work. Maybe Ellen and the others just fraked up the reproductive system of the Significants, whether deliberately or accidentally.

And it's probably a bit of tongue-in-cheek humor from Ronald D. Moore that the show's opening crawl says that Cylons "evolved," when they are actually great creatures created by intelligent design. Much like the "evolution" of piston engines from the Model A Ford to the Maserati.

Raider Cylons

Raider Cylons (to be known in these pages as "Raiders" or "Heavy Raiders") are meat machines. They are near-perfect examples of a cybernetic organism—living tissue, including a brain, interfaced with electromechanical devices.

The machine part of a Raider takes the form of a vehicle able to fly through space as well as through most planetary atmospheres. They are equipped with missiles (either conventional or nuclear) and kinetic energy weapons (a.k.a. bullets) and are employed by the Cylons to attack ships in the Colonial Fleet. They seem to have similar propulsion units as Vipers, with slightly superior maneuverability, though Raiders are also equipped with faster-than-light (FTL) drives for long-distance jumping.

The living part of a Raider is housed inside the metal exoskeleton. The ship's viscera look very much like muscle and sinew, and use oxygen as part of their metabolism. Its brain has limited sentience and thinking ability, so Raiders are about as bright as a

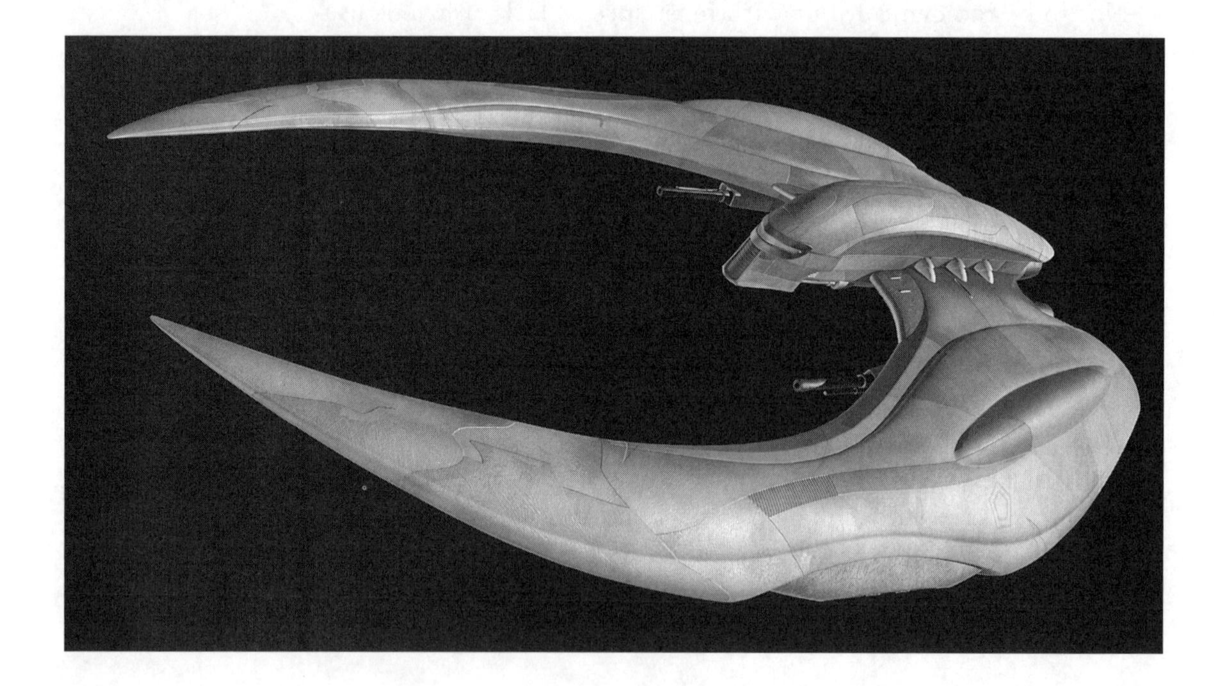

A Raider in flight.

well-trained dog. They can resurrect just like humanoid Cylons—the consciousness of their past incarnations are placed in a new Raider body, and their memories are reborn with them. This gives them the ability to learn from past mistakes, as well as the capacity to develop and hold one hell of a grudge (as seen in the stand-alone second season episode "Scar."). Still, while Raiders are relentless in their motivation, they're not overly bright. Their threat to the Colonial Fleet lies more in their sheer numbers rather than any brilliant tactical skills.

The fourth season episode "He That Believeth in Me" showed an interesting side of Raider cognition. During an attack on the Colonial Fleet, a Raider made some form of visual contact with Samuel Anders's eyeball and recognized his true Cylon nature. That Raider must have immediately broadcast the word to the rest of the Raiders that they were attacking a fellow Cylon, because all the Raiders stopped the attack and returned to their basestar. This "rebellion" of the Raiders caused Cavil to advocate their reprogramming—a position that so revolted some other Cylons that it led to a civil war.

Centurions

Finally, there are the Centurions. Centurions are the soldier-workers of the robot world. They combine tactical knowledge with extreme firepower, making an almost unbeatable combination. We don't know if their tactical knowledge is innate or directed, but it is not too much of a stretch to guess that Centurions are given their overall mission from their controller and are left to their own devices to carry out that mission. They have an outer skin of armor that is impervious to ordinary bullets (though they can be brought down by explosive rounds), and their arms are multiple tools, housing hands, edged weapons, and three-barreled Gatling-type guns.

They have the capacity for sentience, but had been prevented from attaining self-knowledge by the humanoid Cylons, who are appropriately terrified of their own creations rising up and rebelling against

A Cylon Centurion.

them. The Cylons have installed a kludge* called a "telencephalic inhibitor"† that keeps the Centurions dumb and happy. After Cavil's decision to lobotomize the Raiders, rebel Cylons removed the inhibitor from some of their Centurions, giving the Centurions the capacity to turn against Cavil's faction.

Also, as seen in *Razor*, somewhere in the galaxy there still remained a few "classic" Centurions—the original type, probably designed by Graystone Industries around 58BF. They are just as metallic as new Centurions, but they tend to be gold rather than silvery and are much more classically Egyptian in their design. One main difference between classic and modern Centurions is that the classic ones do not have integral weapons. Another is that they can speak.‡ Of the remaining original Centurions, those that weren't destroyed when their basestar blew up at the end of *Razor* were probably destroyed along with the Colony.

*And what a kludge it is! By implication, someone has got to go around installing these modules in every Centurion that comes off the assembly line. Simply changing the Centurion design to remove the capacity for sentience would make a lot more sense, if such a thing were possible. It's like a governor put on an engine during manufacturing to keep the rpms down to safe levels.

†A term that first appeared in the fourth season episode "Six of One."

‡"BY. YOUR. COMMAND."

CHAPTER 3

Are We Creating Our Own Cylons?

Until fairly recently the story of *Frankenstein* was only a metaphor: the label of "crazed scientist who created a monster he couldn't control" could be—and was—applied to any researcher working on nuclear weapons, cybernetics, industrial chemicals, and, starting in the 1960s, genetics. The stakes are even higher now. In the early twenty-first century, scientists are on the threshold of truly *creating* life by the manipulation of laboratory chemicals. The goal is to synthetically replicate the genome of a living creature, or to design a new, viable genome. Some researchers are even trying to physically build the structure of single-celled organisms. This raises the deep, unspoken question that pervades every episode of *Battlestar Galactica*: Exactly what debt, if any, do we owe to the creatures we create?

Genetic research is no longer in the realm of brilliant Ph.D.s holed up in a high-tech laboratory. You can go to your friendly neighborhood toy store and buy a science kit that teaches you how to extract and isolate DNA from nearly any cell. You can

Admiral William Adama.

Admiral Adama, flanked by Lee Adama and Tom Zarek.

go to any Web site specializing in science equipment for high school teachers, and you can buy all the necessary gear to allow you to modify bacteria genetically to give them resistance to ampicillin (many germs are ampicillin-resistant already, so you're not likely to be solely responsible for breeding the next killer plague). Even better, if you're a college student, you can enter MIT's iGEM competition.

iGEM stands for international Genetically Engineered Machine.* The competition is open to teams of college undergraduates from all over the world. The goal is to take public-domain standardized DNA sequences known as BioBricks—for example, the DNA sequence that codes for a specific biological function such as making luminous protein, or for detecting concentrations of various elements—and insert them into bacteria to make useful biological machines that are not evolved, but intelligently designed to perform certain tasks. The comparison with a brick is deliberate—BioBricks are designed to be as easy to use and as interchangeable as LEGO bricks. The process is very much like taking a handful of electronic components—resistors, capacitors, diodes, logic gates—and putting them together in just the right way to build a robot. Of course, life being what it is, there is never any guarantee that the interaction of all the different BioBricks will have the result you expected them to have.

The iGEM program represents a gigantic shift in the way biology is taught. It used to be that the goal of biology was to learn as much as possible about how various life processes work. Now the cool (and scary) thing about synthetic biology is that the goal is to learn how to *use* those biological process as black boxes: you give these processes an input, and they give you an output. There's no need for a synthetic biologist to understand exactly *how* a certain gene sequence codes for a specific protein—all they need to know is that it *does*.

In past years, teams of undergraduate college students from around the globe have taken freely available predefined bits of DNA and created:

1. Bacteria that eat industrial pollutants
2. A biosensor that detects high levels of UV radiation
3. Immunobricks to defeat the bacterium that causes ulcers
4. A bacterium that treats lactose intolerance

*I mean, really—with that name, it's as though they're *asking* you to create Cylons.

5. RNA molecules that encode nearly all of the IF . . . THEN Boolean logic of computers
6. Intestinal bacteria that produce fart gas that smells like wintergreen.

The resulting biological machines may be our first baby steps toward creating Cylons, but so far the iGEM students are still taking BioBricks and inserting them into already living bacteria. They haven't tried to build an entire living organism from catalogue parts. They're leaving that to the big guys like J. Craig Venter.

Venter, the bad-boy genius of the modern genetics industry, was the maverick who wowed the genetics world in the 1990s by boasting that his company would sequence the entire human genome before the federally funded Human Genome Project finished. Ever since then he has been at the forefront in many different types of genetics-related breakthroughs: his latest task was to grow a bacterium from a completely synthesized genome made from parts available in the laboratory. The announcement in May 2010 of his success was hailed as the first creation of totally synthetic life that we know of on Earth.

It's possible that this same type of work—discover the genetic code of an already existing creature, synthesize that code, then grow it in a lab—is what the young pilot Bill Adama stumbled upon just before the end of the first Cylon War. The hybrid on the Cylon ice planet clearly was involved in some form of experimentation: it collected Colonial civilians and probably used sections of their DNA to grow the biological parts that surrounded the hybrid's tank. Perhaps, with the help of the Final Five, those parts became the biological components of Raiders and Centurions. They almost certainly served as the biological components that became the humanoid Cylons themselves.

Frankenstein's creature condemned his creator for building him without a reason. Our biological machines are not in a position to demand accountability from anyone—yet. But research clearly isn't going to stop there, and there may come a time when our creations start to experiment on their creators. At that moment, whether we want to admit it or not, we'll be starting down the same path as the Colonials. When our creations ask "Who am I?" we'll be able to give our Cylons only one answer: "You're us."

CHAPTER 4

Cylon Intelligence and the Society of Mind

What type of brain does a Cylon have? Since Cylons are intelligent organic creatures, virtually indistinguishable from Colonials, their brains must be at least superficially similar to Colonial brains. If you assume that the Colonial brain is similar or identical to a human brain of the twenty-first century, the Cylon brain should meet the following minimum requirements:

- Elements: 100 billion neurons
- Connections: 100 trillion
- Storage: 100 million megabytes
- Image Processing: 210 million images per year
- Processing Speed: 0.001 Megahertz
- Power Requirement: 20 watts

Though it's never specifically stated, a lot of evidence points to the idea that Cylons are considerably smarter than their Colonial counterparts.

Cylon model Two, also known as Leoben Conoy.

Leoben and Starbuck look down at the child she believes is hers.

Six in a projection of a forest.

The most obvious example comes from the episode "Torn," in which Six tells Baltar that Cylons can use their brains to project whatever imagery around them they wish. She gives the example of a corridor in a Cylon basestar: to Baltar it's just a bare metallic hallway, but Six can use the powers of projection within the Cylon mind to make walking down the corridor like taking a stroll through a beautiful sun-dappled forest.

At the very least, Cylons *use* what brains they have differently. One of the most common misconceptions about cognition is that humans use only 10 percent of their brain power. Anyone who has ever seen a PET scan of their own brain activity knows this isn't true—most scans show sporadic activity happening throughout the entire brain. While it is true that not all parts of the brain are active at the same time—when you're sitting quietly listening to instrumental music, you're probably not using the verbal or the motor control portions of your brain—over time, you can rest assured that you use 100 percent of your brain. Cylon mental superiority might just be a result of them using more of their brain at any given time.

If Cylon brains are more powerful than ours, they might have considerably more neurons and synapses, and possibly faster neuronal processing speeds, though that bumps up against the Cylon Indistinguishability Conjecture.* A much more reasonable way that Cylon brains might be different from Colonial brains is in the "software," the ability to perform at a much higher level than Colonials while still maintaining the same hardware.†

A synapse is a small gap between brain neurons, on the order of approximately tens of nanometers, which serves as a junction to bring neurons together. The complex web of connections between neurons is similar to the complex web of connections between transistors in a computer's CPU, with one main difference. Thanks to synapses, neurons are not connected one-to-one; they are connected many-to-many. A synapse allows many neurons to connect to many other neurons by *not really connecting* at all.

When a stimulus to a neuron reaches a certain threshold strength (called the neuron's action potential), the neuron opens several vesicles within itself and releases a chemical known as a neurotransmitter, the purpose of which is to tell the other neurons, "Hey, this neuron has fired!" The neurotransmitter flows into the synapses between neurons and spreads among them, with the closest neuron obviously getting the strongest dose of neurotransmitter. More distant surrounding neurons get considerably smaller doses. Whatever message the firing neuron was trying to send, the neurotransmitter gets the message across. If the amount of neurotransmitter absorbed by any of the surrounding

---

*Which basically says that Cylons are sufficiently like humans that Boomer, Tigh, and Tyrol can endure years of military medical checkups without being outed.

†That additional software might contain functions that would work phenomenally well on our brains, but that we just don't have because they showed no evolutionary advantage. One of the marines watching Six-in-the-Brig wonders how the Cylon doesn't go crazy. The other marine suggests that maybe Six has a way of shutting off parts of her brain. If such ability ever became available to humans, the line to sign up for it would stretch around the block.

neurons is sufficient to reach *their* action potential, those neurons will then fire, reinforcing the original stimulus that caused the first neuron to signal.

Neuroscientists understand that learning takes place when one neuron stimulates another neuron repeatedly and/or continuously for a few seconds or even minutes. When this happens, the "teacher" neuron releases an additional protein that helps the surrounding "student" neurons to grow. When these student neurons grow, they strengthen the connection between themselves and the teacher. Whenever the "teacher" fires off a signal, the "students" will over time become more receptive to receiving it. One stimulus will become tightly associated with a specific response, and the brain will have learned something new.*

One thing brain scientists don't yet fully understand is how a functioning brain—one like ours that can think and reason and create (or at least appreciate) music and art—can be made out of repeated units of a relatively simple object like a neuron. In this way neuroscientists have a lot in common with artificial intelligence researchers who are trying to solve essentially the same problem from the other direction—how to turn a box of computer chips into a functioning brain.

The MIT professor Marvin Minsky is rightly described as one of the fathers of AI. Minsky built the first random neural net computer (the predecessor to Cylons) in 1951 and patented the first head-mounted computer display twelve years later. In the 1970s, Minsky and Seymour Papert started to develop their "Society of Mind" theory. They said that the complex construction we call the "mind"—the realm of language, memory, learning, consciousness, the sense of self, and of free will—is actually built brick-by-brick of much smaller subunits, which he calls agents. These agents are themselves mindless; it is through their interaction that they *produce* the mind.

---

*Incidentally, this offers one theory as to why "smart" people also tend to be more neurotic than most: bad experiences are learned and remembered as equally well as good experiences, and tend to be carried in the brain long after other people would have forgotten them. Remembering bad stuff from the past can have negative influences on decisions and actions a person takes in the present. Sometimes ignorance can be bliss.

As Minsky said in his 1988 book *Society of Mind*: "What magical trick makes us intelligent? The trick is that there is no trick. The power of intelligence stems from our vast diversity, not from any single, perfect principle." In other words, to employ another Minskyism, "Minds are what brains do."

As we mentioned, an important clue that Cylon brains work the same (or in a similar) way came in the episode "Six of One," when we learned about the Centurions' telencephalic inhibitor. The humanoid Cylons wouldn't need that kludge if Centurion brains were a bunch of transistors running Python. Such a gadget would only be necessary if Cylon brains worked as a Society of Mind—if "mind" just *naturally emerged as a consequence* of connecting together a threshold quantity of smaller brain subunits. If that's the case, then twenty-first-century AI researchers here on Earth have only one question: What is that threshold quantity?

Sentience is another word for consciousness, or the ability to be aware of oneself. For example, you know that you are a human.* You know that you are separate from the world around you. When you think a thought or feel an emotion, you know that it only applies to you. You understand that the rest of the world does not change as your mind changes. You are aware that no one else feels or thinks the things you do. You are also aware that your thoughts and feelings come from inside you; they're not placed in your mind from an external source. You have a memory of your thought processes—you know that you didn't always think or feel the way you think or feel right now. And you know that you probably won't think or feel tomorrow the way you do right now.

This self-awareness is one of the foundations of human intelligence. Without it, humans would be nonthinking brutes, unable to plan, or remember, or relate to other people.

Or would we? Is it possible to be intelligent without being self-aware? Social insects, such as bees and ants, display what scientists call collective intelligence. The hive or colony, acting together, has

*Or so you think.

a form of intelligence that no ant or bee could possess on its own. The neural ganglia of an insect are a collection of nerve cells, just about equally divided between the insect's head and abdomen. These neurons act as a simple brain—far too simple to allow the insect to experience anything like complex thoughts. We can safely assume that a single ant or bee has no knowledge of itself as an individual. It literally doesn't have the brain for it. There's no way that one ant could know enough about the colony and the outside world to be an effective administrator.

Yet thousands of ants can, by working together, engage in agriculture, fight battles, build cathedrals, and organize complex foraging expeditions. They do this without any orders from a more intelligent entity. The colony regulates itself. It can adapt itself very easily to changes in the outside environment. The ants do this all without being aware of what they're doing and without control from any organizer at the top.

In the years after World War II, the insect researcher Karl Von Frisch discovered that honeybees perform an incredibly intricate dance after returning to the hive from a successful foraging mission that tells other bees where the food is. Studies have shown that the dance accurately conveys the distance and direction of the food source, which can sometimes be as much as five kilometers from the hive! This nonsentient type of group intelligence is known as emergent intelligence.

When Raiders swarm out of a basestar, or when Centurions use coordinated tactics to execute an attack, are they exhibiting emergent intelligence? Up until "Six of One," the answer would have been a qualified "most likely." There always remains the possibility that Centurions are dumb hunks of metal controlled wirelessly by humanoid Cylons in a nearby basestar—a drone or RPV,* in modern military vernacular. But seeing how they operate, you get the feeling that Centurions and Raiders are like characters in an ultrareal present-day video game. They follow a few simple rules of action (which every game designer likes to call "artificial intelligence"), which can result in amazingly complex behaviors that seem almost "intelligent."

*Remotely piloted vehicle.

Cylon Raiders swarming.

Learning and memory alone are not enough for sentience. Dogs can learn and remember tasks and events. They are marvelously intuitive about the character and emotions of the people around them. When a dog jumps into the air to catch a Frisbee, it is performing a heuristic act that is the equivalent of doing complex mathematical calculations in its head, involving changes in position and velocity. But a dog looking at its reflection in a mirror does not know that it is looking at *itself*. A dog does not "think" of itself as a dog, or indeed as any sort of independent entity apart from the world. A dog thinks (to the extent that a dog thinks) that it *is* the world.*

*This sense of solipsism shows up even more strongly in house cats. Place a small shaving mirror on the floor in front of your cat. You cat will see its reflection and may hiss and posture to scare away the "other" cat. Eventually it will look around the mirror to see the rest of the intruder, and won't find anything. This will drive your cat crazy for a few minutes. When the "other" cat consistently refuses to appear, your cat will lose interest and go to sleep.

On the other hand, a chimpanzee will, after some initial confusion, eventually understand that the ape in the mirror is actually itself, not another ape.* Before long a bright chimpanzee will use the mirrors as grooming tools (much as we do). In other words, chimpanzees, and all the great apes, are almost certainly sentient.

Are Raiders and Centurions more like dogs or chimpanzees? This presents viewers of *Battlestar Galactica* with an interesting problem. In the episode "A Measure of Salvation," we were reluctantly led to the conclusion, along with the leaders of the Colonial Fleet, that killing off all the humanoid Cylons would be genocide. If we accept that all three "species" of Cylon are intelligent to some degree, and that apes are intelligent as well, why do only the humanoid Cylons have the right not to be exterminated en masse? Is it simply easier for us to accept the personhood of other beings the more they look like us?

A 2008 study by scientists at Università degli Studi di Bologna would seem to verify this. You know how, sometimes, if another person is hit or touched, you can actually feel it yourself? Scientists collected pictures of other people having their faces touched by fingers and showed them to test subjects of different ethnic groups. When the person in the photograph was the same ethnic group as the subject, the subjects were more likely to report their own perception of being touched. Then the scientists wanted to see if the phenomenon worked with other groupings. They gathered pictures of left- and right-wing politicians having their faces touched and showed them to the subjects. Once again, more test subjects reported a perception of touch when the observed face belonged to his/her own political group. If we're more likely to share feelings with someone when they're like us, what does this say for our ability to recognize consciousness or sentience in others? What does this say for our own intelligence?

*This was clearly proved in the forehead dot experiment. Ape researchers painted a small dot on the forehead of a chimp while it was asleep, and then presented it with a mirror upon awakening. It took most chimps a few seconds to realize that the reflection was of their own face and that there was a dot painted on their forehead, which they proceeded to wipe away.

CHAPTER 5

How Can Cylons Download Their Memories?

In an article in *Wired* magazine,[1] Kevin Kelly took a look at the computational parameters of the human brain. While the human brain probably can never be compared one-to-one to a computer (at least until computers are built like human brains), there are some similarities. For example, the available evidence seems to indicate that there are approximately 100 billion neurons in the brain, giving rise to 100 trillion synapses. These give the average human brain a data storage capacity of about 100 million megabytes—or 100 terabytes.* This is roughly equivalent to all the printed material in the British Library, New York Public Library, Bodleian Library of Oxford University, and the Library of Congress taken together, or the hard-drive storage of two hundred high-capacity DVRs. If we assume that 100 million megabytes is also the storage capacity of both Colonial and Cylon brains, we can get

*Something clearly pointed out in the rebroadcast of *Caprica*'s pilot episode.

Cylon models Eight and Six.

Cylon models Eight and Six.

Eight awakens in a pool of reanimation goo.

some kind of handle on the most intriguing ability of Cylons—their ability to resurrect.

We first learned about Cylon resurrection in the miniseries, when Six and Leoben respectively told Baltar and Bill Adama that when Cylons die, they wake up in another body. The realization that Cylons could come back to life filled the Colonials with a mixture of fear and wonder. Fear, because the ability to resurrect made the Cylons seem almost like the mythical Hydra—no matter how many heads you cut off, another one would take its place. Wonder, because any sentient creature, aware that they're eventually going to die, is going to be in awe of a creature that has managed to beat death.

We've learned precious little about the actual download and resurrection process since,* but we do know this much: in the episodes

*Moore's law, don't you know.

"Exodus, Parts I and II," it took Brother Cavil approximately two days to resurrect successfully. Knowing that can help us to figure out how the process works.

If Cylon resurrection is roughly equivalent to restoring a computer hard drive by way of a backup, we can assume that most of the duration of a data restore is spent transferring the information from one source to another. Since Cylons don't have to be plugged directly into a Resurrection Ship in order to download their memories, data transfer must be handled wirelessly. For a dead Cylon to transmit 100 terabytes of information over the course of two days requires transmissions of 50 terabytes per day, which is equivalent to 2 terabytes per hour, 34 gigabytes per minute, or 578 megabytes per second. Is such a thing even possible?

As of this writing, commercial wireless data protocols (specifically 802.11n, a.k.a. WiFi-n) offer data transfer rates of about 12 megabytes per second. If Brother Cavil had such a data plan to connect him to the Resurrection Ship, it would take him somewhere in the neighborhood of ninety-seven days to upload the content of his brain. Also, such speedy data protocols like WiFi-n only work over relatively short distances, on the order of a few dozen meters. To send digital data across vast distances through space necessitates using a slower, more reliable data standard. It takes up to seven minutes on a bad day to get a single 12-million-bit image from one of the cameras in orbit around Saturn aboard the *Cassini* spacecraft; downloading an entire brain might take years. On the other hand, it takes approximately fourteen minutes to download a 5-billion-bit image from the much newer, albeit much closer, Mars Reconnaissance Orbiter spacecraft—a 36-fold improvement. So our ability to send vast amounts of data across space is improving rapidly.

To download Brother Cavil in two days, either the Cylons have developed a long-distance data transfer method that sends 45 times as much data as WiFi-n, or they use another method of communication altogether.

One of the most intriguing possibilities for the makeup of Cylon brains is that they may use quantum entanglement.

Quantum entanglement could use a book of its own to explain. The short version is that it is entirely possible to create two subatomic particles in such a way that one particle is intimately linked to the other particle. If the first particle takes on certain attributes, the second particle will take on exactly the same attributes, instantaneously, much like adolescents and fashion accessories. If the Cylons have figured out a way to store brain states in entangled particles, then they technically don't need to upload their consciousness at the moment of death. Everything they think or experience throughout their lives is automatically and simultaneously reflected in a separate Cylon brain on the Resurrection Ship.

Perhaps it takes two days to place a consciousness into a body. If that's the case, then one answer to Cylon backups might be much simpler than quantum entanglement or wide-channel broadcasting. It's entirely possible that Cylons use incremental backups. How much data would a Cylon have to broadcast if it only sent a day's worth of experience at a time? We can make some simple assumptions and come up with a conservative estimate. Let's assume that an average Cylon lives to be a hundred, that the 100-terabyte brain storage capacity will hold an entire lifetime's worth of experience, and we ignore any age-induced memory loss. The Cylon would then live 36,500 days. One day's allocation of that is a "mere" 2.74 gigabytes (or just shy of 22 gigabits). That is less than four and a half times larger than a single high-resolution image from the Mars Reconnaissance Orbiter (MRO). At the rate MRO sends data from Mars, it could transmit that much information in slightly over an hour. So it's reasonable to believe that every night as they sleep, Cylons' thoughts and memories of the day are processed and broadcast to the nearest Cylon basestar or Resurrection Ship. The Cylons would probably interpret this nightly rush of images through their head as dreaming. When they die, as their consciousness fades out, only their thoughts and experiences since their last backup need to be sent. Perhaps, with the proper encoding, it can all be sent before the last Cylon brain cell dies.

CHAPTER 6

A Dialogue between a Smartass Fanboy and a Real Scientist, viz: The "Silica Pathways" into the Cylon Head

Smartass Fanboy: It's so frustrating that Cylons are supposed to be indistinguishable from Colonials, yet they can shove fiber-optic cabling into their arms and interface with a computer!

Real Scientist: Wait a minute. I don't think the Cylons are particularly frustrating at all, given the right context. Remember what Commander Adama said: "Context matters."

SF: But if there's supposed to be no way to tell the difference between a Cylon and a Colonial—

RS: But there is. First, let's recall that Baltar's Cylon detector worked. It just wasn't in his best interest to admit that it did at the time, and he never really had occasion to do so later. His detector wasn't a particularly high-tech form of equipment [which you'll see if you go read "Baltar's Cylon Detector" later in this chapter]. I see his detector as, essentially, a mass spectrometer. He has a detector for silicon, and he sees an anomalously high amount of it.

Athena connects to the Galactica via a fiber-optic cable.

SF: A mass spectrometer?

RS: Just as a regular spectrometer separates light into its component colors and measures the relative intensities, a mass spectrometer takes a sample of a material and determines the sample's chemical composition and relative abundances of different elements.

SF: Okay, but you still haven't answered how the Cylons can connect to a computer!

RS: There is a lot of active research going on, as we speak, on brain/machine interfaces. Let's assume that Cylons are expert at this.

SF: They'd have to be.

RS: Shh. Let us also define what we mean by "silica pathways." Silica has the chemical formula SiO_2, which is the same formula as the mineral quartz or . . . wait for it . . . glass. So when we say "silica pathways," what we are talking about is fiber optics. (I'm shocked nobody picked up on this.) Nerve impulses travel at about 200 mph. Pulses of light travel along fiber-optic cables travel at, well, the speed of light. Okay, slightly slower, since the light is traveling through glass and not a vacuum, but you get the point. If the Cylons could create a microscopic interface between the neurons in their head and their fiber-optic nervous system, that would explain a lot. They would have faster reactions, yet could have biologically indistinct brains when compared to humans, but with different programming. They could go for years without being detected, Sharon could "plug in," they would *still* be nigh-undetectable . . . except with a mass spectrometer.

Cylon model Eight Sharon "Boomer" Valerii.

Saul Tigh, William Adama, Karl "Helo" Agathon, and Sharon "Athena" Agathon.

SF: Ah, but Chief Tyrol has been in the military since he was eighteen. If he's in his early thirties, he's undergone twelve or more years of military physicals without anyone noticing anything strange . . .

RS: . . . which is not inconceivable. Silica is chemically inert—that's why you store acid in glass, that's why beach sand is overwhelmingly quartz (except in places like Hawaii, where the most common local rock, basalt, has little or no quartz). Having had numerous military physicals—including flight physicals—myself, I've never had a test that would find silica in my body. It's simply not a chemical for which they would check. Most physicals check for things like proteins, enzymes, antibodies, sugars . . . all of which could reasonably be within normal human limits for a Cylon. The military does not check for silica because there is no reason to check for it. Nor is there any reason to expect that it would end up in the bloodstream in any way.

SF: Colonel Tigh's military medical record has data going back forty years, and no one has picked up anything weird about him?

RS: Tigh's bloodwork would be expected to be out of the ordinary.

SF: Aha!

RS: It would show anomalously high amounts of glucose and ethanol, but that simply means that he's a high-functioning alcoholic. Nothing would have detected a type of anomaly that would "out" him as a Cylon. It is also not unreasonable to expect that the Final Five were programmed to avoid doctors as much as possible. All part of "The Plan." Further, I think you overestimate the thoroughness of (non-flight) military physicals. If your blood isn't too out of whack and your pee is okay, you're good to go. They simply don't do X-rays, CAT scans, or MRIs as a matter of course.

SF: I'm not done.

RS: Of course you're not.

Saul Tigh: high-functioning alcoholic.

SF: Sam Anders was a professional athlete—he might have been more thoroughly medically scrutinized than the military men, assuming Colonial society had the same problems of athletes and steroid use that our society does!

RS: In all likelihood Anders would have been more highly scrutinized, but it's still unlikely they'd have found him out. He may be considered a physical freak, sorry, outlier, like Lance Armstrong, but there would be no reason to suspect he's a Cylon. Then again, it's not unreasonable to think that his software had "governor" routines so that he didn't stand out too much.

SF: Under those circumstances—under "normal" medical procedures—none of those Cylons would have been detected?

RS: Now you're catching on. Cylons passing routine physicals with their silica pathways going unnoticed is completely possible—given that these glass fibers are much, much thinner than a human hair, given that silica is chemically nonreactive, given that it's not something that doctors would look for, and given that it also does not have a dipole moment, it makes perfect sense.

Sam Anders played professional Pyramid on Caprica.

SF: Dipole moment? I used to know what a dipole moment was . . .

RS: I allude, of course, to MRI scans. An MRI scanner starts by subjecting the object being scanned, usually a person, to an extremely high magnetic field. Water, which composes the bulk of living tissue, reacts strongly and aligns itself with the field. Water does this because it is a polar molecule. There are two hydrogen atoms bonded to one oxygen atom—but the hydrogen atoms have a 108-degree angle between them. So, essentially, water has a "positive" side and a "negative" side—it has "poles" like a magnet. Silica, in its two most common forms, is symmetrical. It does not react to a magnetic field in the same way as does water, hence it would not show up on

BALTAR'S CYLON DETECTOR

We don't yet know the physical makeup of Cylons. Or do we? If Athena can get pregnant by Helo, certainly there exist some major similarities at a fundamental level between their biological makeup and that of humans. If Cylons have silica pathways, then they have a much higher concentration of silicon in their systems, even if at the "parts per million" level, than do Colonials. So how can we analyze the basic makeup of Cylons versus Colonials? What if we could take a tissue sample, then split all the molecules of the sample into constituent atoms, and determine the relative concentrations of the various species of atoms? There's a fairly simple technique called mass spectrometry that Baltar may have used to do just that.

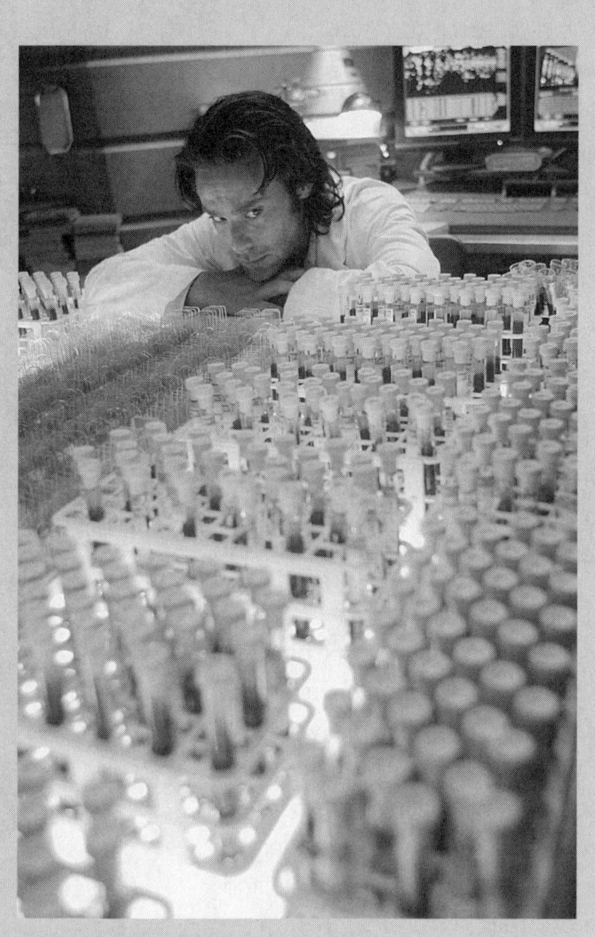

Blood samples in Gaius Baltar's lab.

Mass spectrometers (or simply "mass specs," as scientists often call them) use both electrical charge and mass ratio to separate atoms. If a charged particle is moving within a magnetic field it experiences a force perpendicular to its motion and tends to follow a spiral pattern. If you've ever seen an image of particle tracks in a bubble chamber (a device filled with superheated fluid that is used for detecting charged subatomic particles passing through it), the paths of the charged particles form all manners of curlicues. The radius of curvature—or how tightly wound the spirals are—of a trajectory is related to the charge-to-mass ratio of the particle. The tighter the spiral, the higher the charge-to-mass ratio. The direction of curvature is caused by the particle's charge, positive or negative.

›››

Baltar is the first to discover that Boomer is, in fact, Cylon model Eight.

A mass spectrometer generally has three components. First, the molecules in a sample are split into component atoms and then electrons are stripped off, that is, the atoms are ionized. Next, by accelerating the atoms and applying a magnetic field, the atoms are "sorted" by charge-to-mass ratio. Finally, there are detectors to determine the presence of various elements, even their relative abundances. It is these detectors that would notice a "Cylon spike" in the abundance of silicon, owing to the presence of their internal optic cables.❮

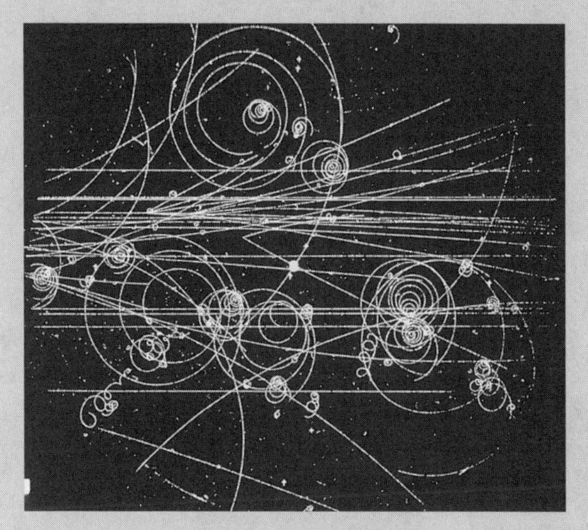

Particle tracks through a bubble chamber.

an MRI. Remember also that the numbered Cylons haven't been in Colonial society that long, maybe a few years tops, so none of them may have ever had reason to be subjected to an MRI scan.

SF: Hang on! Tigh's been Adama's friend for over thirty years! Tyrol's been around for at least twelve! So does that mean that Tigh, Tyrol, Anders, and Tory aren't really Cylons? That can't be, because Anders made eye contact with a Cylon Raider, and Tory has super-strength.

RS: Oh, they are Cylons, they are just fundamentally different from the numbered models. The Ones through Eights were created just after the last Cylon war; the Final Five are a few thousand years old. It's all perfectly self-consistent.

SF: Oh, making them self-consistent is fine and easy. Making them undetectable to a simple X-ray or CAT scan isn't so easy.

RS: Sure it is. First, a CAT scan is basically a three-dimensional X-ray, so for our purposes here, the terms are synonymous. If we assume that the fiber-optic/silica pathways (synonymous) are of the same size scale as nerves, even significantly larger, CAT scans do not have the spatial resolution to see them. More importantly, the opacity of silica in the X-ray portion of the spectrum is very low. Rephrased, the glass of fiber-optic cables is just as transparent to X-rays as is visible light. Neither a CAT scan nor an X-ray would ever see the silica pathways.

Starbuck interrogating Leoben.

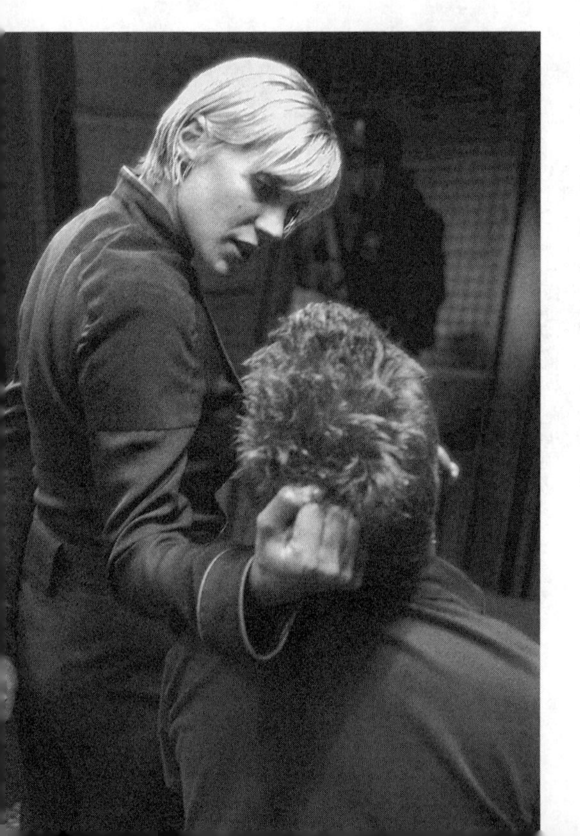

SF: Okay, so Cylons can be distinguished if you put them in a physics lab.

RS: Easily, and this is exactly what happened in "Flesh and Bone." Baltar's Cylon detector worked for Sharon because he was looking for the right thing.

SF: But Cylon neurons have to look and work the same as Colonial neurons, at least as far as a moderately stringent medical test is concerned. If they are different, the Cylon will be instantly found out.

RS: No, they don't have to work the same. The neurons have to communicate with the brain at one end and biological sensors (heat, cold, pain) at the other, and those interfaces would be the biggest technological hurdles in creating skinjobs. There has to be a microscopic transducer of some kind at each end. What happens in between, how the data is sent, doesn't matter.

SF: But that's impossible.

RS: Not impossible, just the biggest technological hurdle, as I said. In fact, back in 1997 scientists at Caltech developed a "neurochip"—a noninvasive device that connects living brain cells to electrodes on a silicon chip. More recently, Infineon Corporation manufactured a chip that allows direct communications between nerves and both sensors and actuators.

SF: Ah, but wouldn't those actuators and neurochips show up on a physical exam?

RS: They would show up on a physics exam. They would add to the silicon signature on Dr. Baltar's Cylon Detector. Given all that we've discussed, I would argue that skinjob Cylons would pass even a very stringent medical exam. You would have to biopsy tissue and look at it under a microscope to see anything amiss. Or place a tissue sample in a mass spectrometer.

SF: A Cylon could be detected via a biopsy, then?

RS: Sure, if any of them had one performed. Tyrol, Anders, and Tory all appear too young and healthy to need biopsies. Tigh just probably says "Get away from me with that frakking needle."

SF: So the Cylons have fiber-optic nerves?

RS: I knew you'd come around. Fiber-optic nerves would explain their faster reflexes (why Anders was so good at Pyramid, why Leoben could move so quickly when being interrogated by Kara). Software (or wetware, the biological equivalent to computer software) in their brains could explain increased strength. Remember, our strength is limited in many ways by pain—or a very similar impulse.

SF: When we learn new stuff, our neurons grow new dendrites to reach out to other neurons, to strengthen the connection that represents the new concept. A prewired network of neuron-to-neuron fiber optics wouldn't allow Cylons to learn anything new.

RS: That's in the head. Cylon brains are biologically similar, with different programming. It's what carries the nerve impulses that's different.

SF: But Admiral Adama very clearly talked about the "silica pathways" into Leoben's head.

RS: Well, that is where they all converge. Besides, in context, when Adama said this it was in the form of an insult, not an analysis.

SF: If all of what you say is true, then it sounds almost like we could build our own skinjobs right now.

RS: It does sound that way, doesn't it?

SF: Don't tell me that skinjobs could be walking among us right now.

RS: By your command.

SF: I won't ask.

RS: Probably for the best.

SF: Okay, next question: Why didn't Cally and Tyrol POP when they got sucked into space in "A Day in the Life"?

RS: You fanboys just do not give up! Besides, that's covered already in chapter thirteen!

SF: Okay, I'll go there . . .

RS: Please do.

CHAPTER 7

Colonial + Cylon + Natives = Human?

Let's define our terms: humans are the people currently living on the planet Earth in the twenty-first century. The people reading this book. The people who watched *Battlestar Galactica* on some form of video display sometime between the Common Era years 2003 and 2009 (2010 if you count *Caprica*). Colonials are non-Cylon people who lived on one of the Twelve Colonies of Kobol around the time of the Cylon attack. Natives are the people living on preindustrial Earth 150,000 years ago.

One underlying but unspoken question throughout the entire show was, "Are the first two groups the same?" Colonials sure look and act human.* Colonials have all the same foibles humans have and no extra powers that humans don't—they get sick and die of cancer, and they can't resurrect. So humans and Colonials are the same, right?

*But then again, so do humanoid Cylons.

Not exactly. There have long been clues that something fishy was going on regarding the two species. We got our first clue that humans and Colonials were distinct even before Hera Agathon was born. In the second season episode "Epiphanies," Dr. Cottle, having examined the fetus in sickbay, said that Hera's fetal blood was "damned odd." Baltar followed by saying that the fetus has "no antigens. It has no blood type."

Hera has no blood type. This fact astonishes Baltar and Cottle, and it even sounds pretty science fictiony: A person *with no blood type? Cool!*

Settle down. There's a decent chance that *you* have no blood type, either.

In France in the 1600s, Jean-Baptiste Denys, King Louis XIV's court physician, made a name for himself by transfusing lamb's blood into several people in an attempt to save their lives. Unfortunately, the name he made was "the defendant": after a few seemingly successful transfusions, Denys was arrested when one of his patients died. He was tried and found not guilty, but—much as our own government has forbidden certain types of stem cell research—the court forbade all kinds of transfusions from then on.

The subject of transfusions lay more or less dormant until 1901, when the researcher Karl Landsteiner discovered that human blood has three main antigen patterns: either one type of antigen (which Landsteiner called "Type A"), a different type of antigen ("Type B"), or *no antigens at all*. Landsteiner called this "type 0" to indicate zero antigens, and that quickly became corrupted to "Type O." Two years later, other scientists discovered blood with both A and B antigens, a much more scarce type known as Type AB.

It so happens that here on Earth, Type O is the most common blood group, possessed by nearly two thirds of all humans. The complete lack of antigens means that in a pinch, Type O blood cells can be transfused into people with Type A, B, or AB blood without causing an antigen reaction. Without Type O blood, blood banks and transfusions would be a lot more difficult to manage, and would be short-stocked far more often.

Lee "Apollo" Adama.

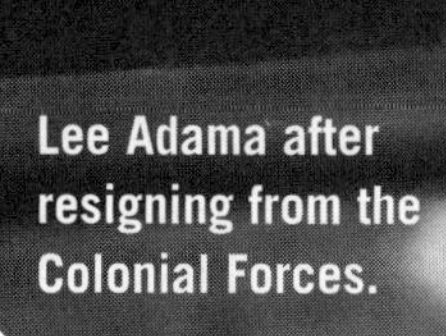

Lee Adama after resigning from the Colonial Forces.

The Colonial population knows *nothing* about this. We don't know what blood types they have— they might all be Type A, or Type B, or a mix of both, or their blood antigens might be totally different from ours. We do know that based on Baltar and Cottle's reaction, they've never heard of Type O blood.

For blood type, human beings get a specific gene sequence from each parent on chromosome 9. One particular sequence of DNA codes for Type A blood. Another, slightly different sequence of codes for Type B. What is most likely a genetic mistake codes for Type O.

This probably means that somewhere in the ancestry of human beings, some people were born with a genetic error in their DNA.* They should have had the sequence for Type A blood, but somehow their DNA had deleted the code for a single amino acid, guanine. The resulting protein was meaningless: it didn't make Type A blood; in fact it didn't do anything. Without a complete Type A protein, these people had no blood type.

A child's blood type can be determined from the genes they get from their parents. Blood types A and B are dominant over Type O, so if a child got an A gene from the father and an O gene from the mother, the child would be Type A. It works like this:

	Father A	**Father B**	**Father O**
Mother A	Type A	Type AB	Type A
Mother B	Type AB	Type B	Type B
Mother O	Type A	Type B	Type O

If the alleles for A, B, and O were evenly distributed throughout the population, we would see blood types A, B, AB, and O appear in a 3:3:2:1 ratio. Type O would be about 11 percent of the population, and types A and B together would make up about 66 percent of the population. But in real life those numbers are nearly reversed. Humans must be heavily weighted with Type O alleles.

Again, this is all new to Colonial science. If they had a Type O allele in the Colonial genome, no matter how rare it was, at some point in their history two Type O carriers would have met and had a child without blood antigens. For Hera's blood to be a total anomaly,

*Some studies indicate that this might have happened as many as three times in the course of history.

the Type O allele must not exist on the Twelve Colonies. Is this proof that the Colonials and humans are different species? Not exactly, but it does show a curious separation between the populations: Humans could be Colonials with a strange genetic alteration, or Colonials could be a very highly restricted group of humans.

Our next clue came in the season three episode "A Measure of Salvation," in which the Colonials find a beacon from the thirteenth colony, as well as a room full of dying Cylons onboard a Cylon Baseship. Dr. Cottle says that the Cylons are sick with lymphocytic encephalitis. He says it is spread by rodents, which is perfectly true, and that "humans" developed an immunity to it—which is most emphatically *not* true.

At least, it's not true on Earth. Lymphocytic encephalitis is a real disease, caused by the lymphocytic choriomeningitis virus. In the United States, you're most likely to get it from breathing in the dried urine of a common house mouse. The Centers for Disease Control figures that between 2 and 5 percent of the U.S. population have been infected by the virus, and while most people don't get sick, the ones who do are in for quite a ride. Symptoms include fever, malaise, lack of appetite, muscle aches, headache, nausea, and vomiting. If you're lucky, you won't get the other, less frequent symptoms: sore throat, cough, joint pain, chest pain, testicular pain,* and parotid (salivary gland) pain. But that's not all. Just as you start to feel better, the second phase of the disease hits a few days later: drowsiness, confusion, sensory disturbances, and/or motor abnormalities such as paralysis. In really bad luck cases, the virus also causes acute hydrocephalus (increased fluid on the brain), which requires surgery, and/or inflammation of the spinal cord. The good news is that only about 1 percent of patients die.

It certainly doesn't sound like "immunity."

Some of you are already seeing where this leads: Colonials are completely immune to the lymphocytic encephalitis virus. Humanoid Cylons are killed by it. Earth "humans" are sickened, but very rarely killed.

---

*Where available.

We had enough evidence by the third season of the show to surmise that Earth humans are somewhere between Colonials and Cylons.

But come on, wouldn't we know if we were Colonial-Cylon half-breeds? Not necessarily. Historically, it has been very difficult to determine exactly what a species is. More precisely, it's been very difficult to determine exactly what species an organism belongs to. It is also difficult to recognize when the new species has branched off from an older species in the wild.

As far-fetched as the idea that we are Colonial-Cylon half-breeds sounds, it does have a possible historical precedent. One way to look at the species relationship between Cylons and Colonials might be to look at a similar relationship: the one between modern-day humans and Neanderthals.

Neanderthals were human beings who lived in caves across Europe around 130,000 to 30,000 years ago. They have been used as the stereotype of "cavemen," but they really weren't. They were an intelligent and musical band of people who probably had some form of religious beliefs and a detailed social structure.

Were Neanderthals the same species as modern humans? For years, the general scientific consensus was that they were not: *Homo sapiens* was of a slightly different lineage from *Homo neanderthalensis*. Then for years it was believed that they were: we were *Homo sapiens sapiens* and they were *Homo sapiens neanderthanensis*, a subspecies. Now current thinking is swinging back to a separate identity for the Neanderthal. But the question still remains: what caused the Neanderthal, whatever they were, to vanish from the scene about 30,000 years ago? There are three main hypotheses:

1. The Neanderthal evolved directly into modern-day humans.
2. The Neanderthal were replaced, probably violently, by modern-day humans (they were killed off or died out).
3. The Neanderthal interbred with modern-day humans and were gradually subsumed into a hybrid population.

The likelihood that the Neanderthal evolved directly into humans is a very old and now almost universally discredited notion, formed back in the days when we thought these cavemen were our direct ancestors. Preliminary results of the Neanderthal Genome Project show that while modern humans and Neanderthal share 99.5 percent of their genome, the two species firmly split and separated from a common ancestor about 400,000 years ago. It is next to impossible that the Neanderthal evolved directly into us.

The second hypothesis is much more interesting: Modern-day humans evolved out of Africa, moved into Europe, and wiped out the Neanderthal in what is essentially genocide. Evidence is starting to mount for this explanation. Excavations in Shanidar, a cave located in the Zagros Mountains in Iraq, from 1957 to 1961 yielded nine sets of Neanderthal remains. Recent analyses of the remains from the individual known as Shanidar 3 at Duke University point very strongly to the conclusion that this adult male was killed by a thrown spear—a weapon that Cro-Magnons possessed, but Neanderthals did not.

Genocide also seems to be the chosen plan for Cylons to destroy Colonials, but do they really have to resort to that? Throughout the show, as the Fleet's population dwindled it was quite possible that the Colonials would find themselves in a genetic bottleneck, unable to maintain a viable population. Baltar himself calculated that the Colonials had eighteen years left before they went extinct. In that case all the Cylons need do is to wait for the last Colonial to die off. If they are computer-based or computer-like beings, waiting shouldn't be a problem for them.

Perhaps that's what happened to the Neanderthals. It wasn't that more modern humans came in and wiped them out through war or disease, but simply that modern humans, having better evolutionary advantages and being better competitors for limited resources, may have simply outlasted the Neanderthals. Life is the survival of the fittest, and compared to modern humans, Neanderthals were not the fittest.

The British scientist Stan Gooch, in his 1979 book *Guardians of the Ancient Wisdom*, hypothesized a third possibility: that Cro-Magnons, essentially an earlier version of modern humans, evolved out of Africa and settled in northern India, while at the same time earlier tribes of the Neanderthal were evolving in Europe. Around 35,000 years ago

MITOCHONDRIAL EVE

In the last scene of the last episode of *Battlestar Galactica*, Angel Six and Angel Baltar appear behind a bearded man (Ron Moore, in a goodbye cameo) at a New York newsstand, reading an issue of *National Geographic* magazine over his shoulder. Angel Six, in voiceover, reads, "Mitochondrial Eve is the name scientists have given to the most recent common ancestor for all human beings now living on Earth." We're supposed to assume they're referring to Hera, Helo and Athena's half-Colonial, half-Cylon daughter.

Lords of Kobol! An actual admitted scientific mistake!

The problem arises from the conflation of two very different terms: "Mitochondrial Eve," and "most recent common ancestor." The Most Recent Common Ancestor is, as the name suggests, the most recent common ancestor of all humans alive on this planet. Mitochondrial Eve is the most recent common ancestor of all humans *along the matrilineal line*. And to explain the difference—as always—we need a little background.

Biological cells come in two main types: prokaryotes and eukaryotes. For our purposes, the most important difference is that eukaryotes have a distinct nucleus and distinct organelles with their own membranes, while prokaryotes generally have neither. Mitochondria are tiny organelles embedded deep in the cytoplasm of almost every cell in your body. They have been called biological batteries or powerhouses because their chief task is to convert glucose into adenosine triphosphate (ATP), the energy unit of cellular metabolism.

Though mitochondria are embedded in your cells, they are self-contained entities, very similar to prokaryotes. For this reason, the biologist Lynn Margulis suggested in 1966 that billions of years ago, primitive mitochondria actually *were* prokaryotes that entered into a symbiotic* relationship with other cells. That hypothesis was reinforced in the 1980s when researchers showed that mitochondria have their own set of DNA, different from their parent cell.

Geneticists almost immediately realized that DNA from the mitochondria (mtDNA) could help them to track evolution and heredity along the female line. Since sperm do not contribute mitochondria† to the developing embryo, an analysis of mtDNA can help to track matrilineal descent

**Sym* = together, *bio* = life; a symbiotic relationship is one in which at least two different organisms live together, usually (but not always) providing a benefit to both creatures. For example, we provide cats with shelter and a steady food supply; cats provide us with aloof companionship and rodent control.

†Specifically, the sperm's mitochondria are marked for elimination by the egg's cytoplasmic destruction machinery. Talk about being a ballbuster.

through the use of specific DNA markers. And because mtDNA isn't repaired as efficiently as nuclear DNA, it mutates approximately ten times faster.

Since mtDNA comes only from the mother, you will have that same code sequence in your DNA; if you are female, you'll pass that code sequence to your children. If you happen to have a mutation to your mtDNA, you'll pass that mutation, which will be shared with all of your subsequent descendants. By tracking layers of mutations backward, geneticists can determine which populations are ancestors to which other populations.

"Mitochondrial Eve" is the term given to the woman who was the matrilineal most recent common ancestor for all humans living on planet Earth today. Passed down from mother to offspring, the mitochondrial DNA of every human is directly descended from hers. Although they lived thousands of years apart, Mitochondrial Eve has a male counterpart in Y-chromosomal Adam, the patrilineal most recent common ancestor. By tracking mtDNA mutations, scientists have determined that Mitochondrial Eve lived approximately 170,000 years ago,* give or take a few tens of thousands of years. She most likely lived in East Africa,† when modern *Homo sapiens* was branching off as a species distinct from other humans.

It's important to emphasize that Mitochondrial Eve and her contemporaries had offspring, and those offspring had other offspring. But throughout the subsequent generations, for one reason or another, the lineages of Eve's contemporaries all died out. Of all the women alive then (and in our case, that means the entire female population of *Galactica* and the Fleet), only one has offspring alive today. We know her as Hera Agathon.‡

This does *not necessarily* mean that Hera is our Most Recent Common Ancestor (MRCA). Hera populated today's Earth solely through her daughters and daughters' daughters. The MRCA is the person who, while no doubt descended from Hera, populated today's Earth via their daughters *and/or sons*. By adding males to the mix, the MRCA almost certainly cannot be the same as Mitochondrial Eve. In fact, most researchers today feel that the MRCA lived only about five thousand years ago, 145,000 years after Hera.

Does such a recent MRCA imply that the human race was once nearly wiped

---

*Incidentally, the use of the term "Eve" is *not* meant to indicate that she was the only human female of the time, simply that she was mother to us all.

†The article that Angel Six reads over Ron Moore's shoulder purports to say Tanzania.

‡Or do we? Head Baltar gleefully said, "Along with her Cylon mother . . . and human father!" It's totally reasonable to think that once the Fleet settled on Earth, other Cylon-Colonial pairs had offspring in the same generation as Hera, perhaps even Six and Baltar. Imagine if *their* child was Mitochondrial Eve!

out, where it had to bring itself back with a small number of survivors after almost going extinct? Not necessarily. If cousins mate with each other, as has been known to happen in tightly knit tribal societies, then the number of ancestors each person could have would be constrained. In some societies, even more recent MRCAs are possible.

Is Hera Agathon Mitochondrial Eve?

There was a real population bottleneck in our history. It took place seventy-five thousand years ago and was called the Toba catastrophe.

Somewhere in our deep past, a giant volcanic eruption—most probably of Mt. Toba on the island of Sumatra—created the volcanic version of a nuclear winter. The supercolossal explosion was the equivalent of one trillion tons of TNT, and sent volcanic debris more than twenty-five miles into the stratosphere. The resulting ash cloud covered much of the world, causing temperatures to drop as much as 5 degrees and possibly triggering an ice age. The

number of humans, already relatively small, dwindled to approximately fifteen thousand, spread throughout Africa and southwest Asia. Yet those fifteen thousand managed to regroup and repopulate Africa within a few tens of thousands of years, and to move out into the rest of the world thirty thousand years later.

Population biologists talk of something called a minimum viable population, which is the smallest number of individuals that can survive "in the wild." For terrestrial vertebrates, that number is around four thousand. Of course, more individuals are always better for the species, as long as the food holds out, because they bring more genetic diversity into the population. As long as there are at least four thousand souls in a single population group, then Hera's children should have survived.❮

the two species met up in the Middle East and Southern Europe, mixing and mating and forming a hybrid population—a cross between Cro-Magnons and Neanderthals. Eventually, by 15,000 years ago, Gooch states, pure Neanderthal and pure Cro-Magnons had died out, replaced by a hybrid of the two—essentially us, modern humans, who combine the best of both species. There is evidence both for and against this hypothesis in the preliminary results of the Neanderthal Genome Project.

Perhaps the final scenes of the final episode show that something like Gooch's theory was played out: Colonials and Cylons interbred with each other and natives they found on Earth. In the process, Hera Agathon and her siblings and offspring combined the best of three worlds to make us what we are.

CHAPTER 8

The Colonial Pharmacopeia

For a ship that was about to be decommissioned, *Galactica* had a remarkably well-stocked pharmacy.

Bittamucin

Revealed in the episode "The Woman King" as a cure for Mellorak Sickness, bittamucin sounds somewhat like a twist on the common antibiotic names streptomycin or erythromycin. But not so fast. It turns out that bittamucin's root is not *-mycin* ("from fungus"), but *-mucin* ("from mucus")

Mucus???? Snot??? As *medicine*!?!?

Yes. It's hard to believe when you're suffering with a cold or seasonal allergies, but mucus is actually your friend. Its slippery wetness keeps your

Dr. Michael Robert and Helo in "The Woman King."

mouth and nasal passages from drying out, which otherwise would make them more vulnerable to infection. If you do get sick, mucus's built-in antiseptics and immunoglobulins work to surround, kill, and remove hostile bacteria from your body. On the healthiest day of your life, you produced about one liter of mucus—when you were sick, you produced much, much more.

Mucins, the key components of mucus, are giant (though still microscopic) proteins, coated with sugar molecules and water.

Ordinarily, such wet, sugary proteins band together and become wet, slimy mucus, but two special proteins—immunoglobulin A and immunoglobulin D, both antibodies—are also sugar-coated, and are released into the bloodstream to fight disease.

When we say a disease is spread via airborne transmission, it is actually spread on small droplets of mucus that we spray out during a sneeze. Could those drops be killing some germs while carrying others?

The answer seems to be a very qualified yes. Although nearly half of Earth's human population carry the bacterium that causes stomach ulcers, very few people, relatively speaking, actually develop ulcers. In 2004, scientists in Japan discovered that some stomach mucins can

Number Six with Gaius Baltar.

Scientist Dr. Gaius Baltar.

have an antibiotic effect against *Helicobacter pylori*, the ulcer bug, stopping *Helicobacter* from constructing cell membranes, effectively letting the bacteria leak out all over the place before it can do any damage. Our civilization is at the stage where we're still investigating the use of mucins as antibiotics. In a few years, perhaps some drug company will take things a step further and actually try to build antibiotics out of snot. It's not too difficult to expect that the Colonial civilization also discovered this fact, and created an entire line of glycoprotein-based antibiotics, all based around the suffix -mucin.

Morpha

It's a pretty good guess that *Galactica*'s painkilling drug morpha is related to our painkilling drug morphine; for a while after it was isolated from opium in 1804, morphine was in fact called morpha, after Morpheus, the Greek god of sleep.

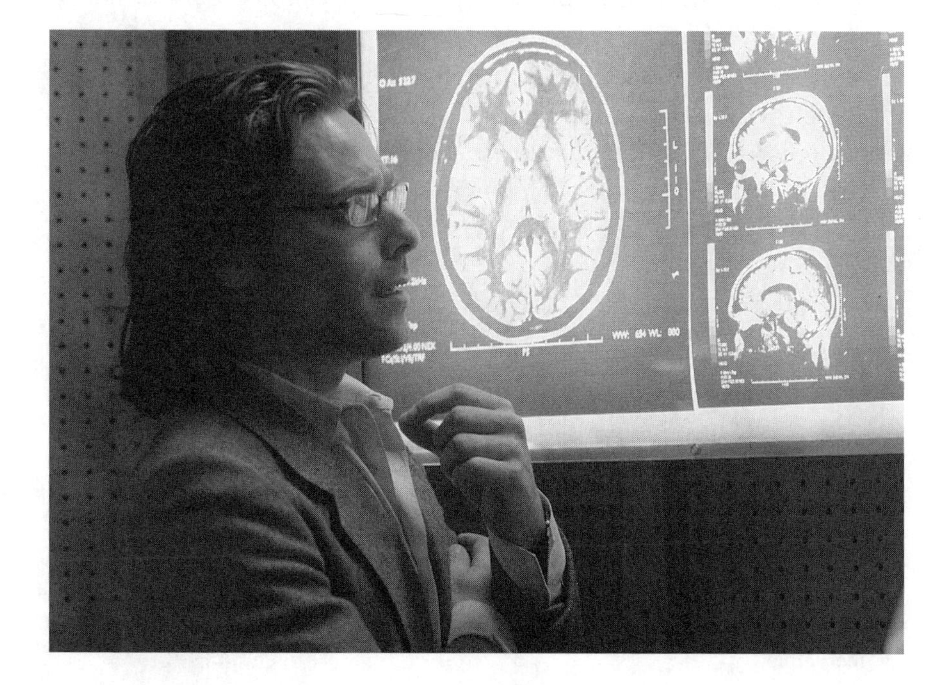

Gaius Baltar in front of brain scans.

Morphine is generally regarded by the medical community as the gold standard of painkilling compounds. In pharmaceutical literature, newly developed painkillers are commonly rated as having some multiple (or some fraction) of morphine's analgesic ability.*

Morphine works by binding to special receptors in the brain that are always on the lookout for opium-like chemicals. Most of the time, those chemicals are released by your own brain, either to fight pain or to produce happiness. The good feeling you get from acupuncture, eating, strenuous exercise, and orgasm are all caused by naturally produced chemicals called endorphins that act like opium compounds upon the brain. If you're in any kind of mild pain or if you just ate or had sex,† there's a good chance you're slightly doped on a close cousin to morphine coursing around your brain right now.

Morphine is extra-dangerous because it fits the brain's receptors especially well; not only does it make you feel good, it knocks away your natural endorphins so that after a short period of use, you *can't feel good* without morphine. The resulting physical addiction can, with care, be overcome, but morphine's psychological addiction can last a lifetime.

Morphine also has a depressant effect on the respiratory system—too much will literally stop your breathing. As we saw in the second season episode "Valley of Darkness," when Tyrol euthanized Socinus on Kobol, this makes it possible to kill someone (accidentally or deliberately) with an overdose of morpha.

Moxipan

In the episode "The Road Less Traveled," Specialist Galen Tyrol wonders why his wife Cally committed suicide—after all, wasn't she taking the antidepressant Moxipan? His unspoken rationalization is

*Not that you're likely to see an advertisement claiming a new drug is "Twice as powerful as morphine!"

†Or if you are really, *really* enjoying this book.

that only depressed people commit suicide, and antidepressants are supposed to end depression, so what gives?

Unfortunately, Tyrol's agony and confusion—and his misunderstanding—are not all that uncommon here on twenty-first-century Earth, either. While antidepressants can be wonderful drugs, giving millions of people a new lease on life, in a small number of cases the side effects of certain drugs can literally become too much to bear.

Most antidepressants work by changing the operation of neurotransmitters, the signaling chemicals of the brain. Some of the most popular antidepressants of our time are known as SSRIs: selective seratonin reuptake inhibitors. When the neurons in your brain fire, one of the neurotransmitters they release is seratonin. Seratonin induces other nearby neurons to fire, and is then reabsorbed by the brain. SSRIs work by slowing down the reabsorption of seratonin. Researchers still aren't precisely sure why this relieves depression, but apparently leaving a small amount of seratonin between the synapses improves the sending of nerve impulses—and improves mood.

Everything has a cost, however. For some people, the relief SSRIs provide comes with side effects varying from weight gain to weight loss, and from mania to a complete loss of all emotional display. In very rare cases, SSRIs bring a deeper depression than existed before. In October 2004, the FDA instructed SSRI manufacturers to include a "black box" warning in their packaging, informing doctors and patients that the medication can increase the risk of suicidal thoughts, ideation, and behavior in children and adolescents up to the age of twenty-five.

Cally certainly seemed to be around twenty-five years of age or even younger. While it's likely that she was sent over the edge by the discovery that she was married to a frakking Cylon, it's not impossible that the SSRIs contributed to her suicidal actions.

Stims

Humans and Colonials get fatigued. We expend energy on various tasks, both physical and mental, and afterward we need to rest to restore various chemical balances in our body. It's part of being human.

Sometimes, however, we can't take a break. The rhythm of society is such that it is unusual, possibly even career suicide, to take a nap during the workday. Members of the military, from Neanderthal raiding parties to twenty-first-century soldiers, may be asked to perform beyond the limits of their endurance. *Battlestar Galactica* is infused with the idea that letting down one's guard, even for an instant, can literally mean the destruction of your entire species.

In times like these, people may turn to stimulants—drugs that increase nervous system activity. Stimulants artificially make it easier to draw on reserves of strength, and thus keep working long after it would normally feel exhausted. The down sides are that (1) spending one's reserves almost always makes it more difficult to recover, and (2) this kind of artificial stimulation is *incredibly* addicting.

Throughout history, stimulants like caffeine, cocaine, theobromine,* and nicotine have been obtained from plants. These stimulants generally work by blocking the reabsorption of one or another neurotransmitter in the brain.† Newer stimulants like amphetamines actually increase the amount of norepinephrine, dopamine, and seratonin‡ in your brain. In addition to zapping the nervous system, both groups of stimulants also pump up the heart rate, dilate airways in the lungs, bring blood to the muscles, and generally prepare the body for action. Overdoses can bring about a dangerously rapid heart rate, a chronically dry mouth, uncontrollable movements and convulsions, and acne. Sometimes an overdose can lead to a complete heart storm, in which the heart quivers randomly without maintaining a steady beat. It's a bizarre feeling, and probably your last.

---

*Contrary to popular belief, chocolate does not contain caffeine; it contains a close chemical relative—theobromine.

†Cocaine blocks the reuptake of dopamine, caffeine blocks the uptake of adenosine, and nicotine blocks the reuptake of acetylcholine in the brain. With all those excess neurotransmitters floating around your synapses keeping your neurons firing away, it would be surprising if you *didn't* feel tweaked.

‡This makes sense, given what we have learned about antidepressants. SSRIs keep more seratonin between your synapses, bringing your mood from sad to okay. Amphetamines bring even more seratonin to your synapses, bringing your mood from okay to *great*!

We've seen in episodes "33," "The Passage," and "Final Cut" that stimulants—and the trouble they cause—are made available (sometimes under orders) to Viper pilots as needed. We saw in the episode "Final Cut" that stims can be acquired and abused by Viper pilots (to wit: Kat) who try hard enough.

Serisone

We commonly hear that our bodies are about 70 percent water, but we don't usually think about where that water is. Our blood is obviously watery, and so are spit, mucus, tears, and urine. Our muscles, fat, and most organs are loaded with water. The vitreous humor of our eyeballs is 99 percent water. Our bones, while still in our bodies, are amazingly juicy—the phrase "dry as a bone" should really be "dry as a dead bone."

The one place you don't want water is in your lungs. There's plenty of moisture there already—oxygen gets into your bloodstream by dissolving in a very thin layer of phospholipid and protein fluid in the alveoli of your lungs, which then transfer the oxygen to your bloodstream. Any additional fluid in your lungs creates a problem—the additional fluid might very well absorb more oxygen, but will also make it more difficult for the oxygen to reach your capillaries.

In the episode "Scattered," Socinus is injured in a Raptor crash on Kobol and has trouble breathing, either from trauma or from smoke inhalation. His comrades inject him with serisone, which quickly eases his breathing discomfort. Based on Socinus's injuries and the rapid action of the drug, serisone is probably a diuretic, a drug that increases urinary output, akin to our own drug furosemide.

Furosemide and similar diuretics work by blocking the reabsorption of sodium, potassium, and calcium in the kidneys. By keeping more of these chemicals in the bloodstream, furosemide tricks the body into thinking it has too much water and needs to get rid of some. Your body sends out an alert to dump all excess H_2O overboard in the form of urine; in Socinus's case, that will relieve some of the fluid building up in his lungs.

THE COLONIAL GUIDE TO GENETICS

When President Roslin was nearing the "endgame" in her fight with cancer, it was Hera's blood—more likely the DNA within her blood's cells—that gave Laura two years of extra life. The database of life—specifically, how to make the proteins that perform the work of a cell—is carried within DNA.

DNA is deoxyribonucleic acid, a molecule that contains four smaller molecules called bases, held together by a sugar and phosphate backbone, and constructed into a double-stranded molecule, like a twisted ladder. The four bases are the chemicals guanine (G), cytosine (C), adenine (A), and thymine (T). They pair up across the two arms of the DNA molecule in predictable ways—G is always paired with C, and T is always paired with A;* hence these are called *base pairs*. The pattern in which these bases appear along the DNA strand—GCCATGGTAGTCAGT, etc.—is the

Laura Roslin and infant Hera.

*In rare cases, the pairings don't match: G pairs with A, T, or another G; T pairs with C, G, or another T, and so on. This causes no end of genetic problems. These mistakes are totally natural, and can usually be fixed by the body's own repair mechanism. Usually. Sometimes, if the body is exposed to a toxic chemical or to certain forms of radiation, the error rate is increased to the point that the repair mechanism is overwhelmed and the genetic mistake remains. Sometimes this expresses itself as cancer, sometimes as a harmful mutation. Very rarely, it serves as a helpful mutation.

›››

»»

information that controls exactly which proteins are made and in which sequence. DNA is a huge molecule—about three billion base pairs long—so it is not only a database, it is a very large database.

DNA is a template for making amino acid chains, or proteins. The pattern GCG, for example, corresponds to the amino acid alanine; ACG is threonine, and so on.

So how does Hera's blood "know" to put Laura Roslin's cancer into remission? Within your DNA are specific sections, like individual records in a database, called genes. A portion of each gene, the encoding sequence, determines what the gene does—its *trait*—and a portion determines whether the gene is active, or *expressed*. When parents produce offspring, the child inherits some gene sequences from both parents. So Hera received some of her traits from her Colonial father and some traits from her Cylon mother.

A research team led by Ehud Shapiro of Israel's Weizmann Institute of Science has developed a new mixture of enzymes and DNA that can combat cancer. By using the data storage and chemical synthesis power of DNA as a biochemical pharmacy, the treatment is designed to detect the chemical markers left behind by cancer cells and respond by producing the necessary drugs. So if the Cylons put genes that encode for cancer-fighting proteins within the genetic sequence of the Sharon model, *and* if Athena passed that trait to Hera, then it's not unreasonable that Hera's blood could have given Laura Roslin some extra time. In the universe according to *Battlestar Galactica*, we all share Hera's mitochondrial DNA (see chapter 7, "Mitochondrial Eve"), and presumably some of her nuclear DNA as well. But somewhere along the way new cancers developed, or in the mix and shuffle of the generations we lost the cancer-fighting gene altogether, because we don't seem to have that Cylon immunity anymore.‹

Interrogation Drugs

In vino, veritas.*
—Roman proverb

"Taking a Break from All Your Worries" is one of the darkest third-season episodes. In it we get to see another side of William Adama, and we learn that the military man who fought hard for civil rights is also an experienced

*"In wine, truth."

interrogator, well-versed in the application of truth drugs on unwilling detainees.

The oldest "truth drug," ethyl alcohol, was known and enjoyed by prehistoric humans and all early cultures from China to Egypt to Mesoamerica. It was seen as a relaxant and an intoxicant, and it was used by shamans to induce visions that would enable them to see the "truth" of a given situation. In Republican Rome, people testifying before the Senate were given a cup of watered wine—just enough to loosen their tongues and make them speak more openly, but not enough to get them so tanked that they made stuff up.

As pharmaceutical knowledge grew, other so-called truth drugs entered the canon. Sodium thiopental and scopolamine* made their way into detective novels, spy thrillers, and more than a few real-life police stations as chemical keys to unlock a suspect's mouth. Both drugs are sedatives and force suspects to lower their guard, but there is no guarantee that what they are babbling is the truth.

In the 1950s and 1960s, the United States' CIA ran a program called Project MK/ULTRA that was designed, among other things, to find the perfect interrogation drug. For a while the CIA was in bureaucratic love with a new compound developed in Switzerland in 1943. Called lysergic acid diethylamide, it would come to be known around the world as LSD. The CIA figured that an unexpected "bad trip" on LSD would make a person say anything, even divulge their carefully hidden truth, to get out of their personal hell. This appears to be the method that Adama uses on Baltar, and he quickly discovers what the CIA learned forty years ago—hallucinogenics might indeed make a person talk, but the results are too unpredictable, and sometimes the drugs make a person feel invincible. Especially when used on brilliant sociopaths (like Gaius Baltar).

---

*Scopolamine also can cause short-term amnesia. It was commonly given to women in labor not as a painkiller, but as a pain *forgetter*, in the expectation that when the drug wore off they wouldn't remember how horrific labor had been. If you were born in the United States between roughly 1940 and 1970, there's a good chance your mother was looped on scopolamine at the time.

Anti-Radiation Medication

Throughout a good portion of the first season, Helo is scrapping for himself on a nuked Caprica. When he's not looking for food, dodging Centurions, or impregnating Cylon Number Eight (known as Caprica Sharon, C-Sharon, or Boomer), every so often we see him stick a hypodermic into his neck from a kit labeled "Anti-radiation." What is he doing and how does it help?

One dangerous component of nuclear fallout is the isotope iodine-131. Just like regular iodine, iodine-131 concentrates in the thyroid, which is bad because you really don't want radioactive material to concentrate *anywhere*. Helo is almost certainly injecting saturated concentrations of potassium iodide into his neck, thereby concentrating "good" iodine in his thyroid. If he subsequently eats or breathes in any radioactive iodine, it will find that there's no more room in his thyroid, and will be excreted from his body.

Another anti-radiation medicine he might be injecting is iron ferrocyanide. Many radioactive by-products of nuclear detonations are heavy metals, an ill-defined group of elements that, at best, can be defined as toxic, whether or not they are radioactive. Iron ferrocyanide has the ability to bind to positively charged heavy metal ions and safely excrete them out of the body.

Bloodstopper

It's yet another example of the bad luck of Felix Gaeta. As temperatures and tempers rise on *Demetrius*, two tall, broad-chested, athletic alpha males get into a dispute with guns. Who gets shot by accident? The little brainy guy.

At least he wasn't ignored. Immediately after Gaeta was shot, Starbuck called for a first aid kit and something called "bloodstopper." We see her take a small sachet out of the box, rip it open, and pour a granulated powder onto Gaeta's wound. Gaeta screams, but we're also pretty sure he stops bleeding. What exactly is in Starbuck's magic powder?

Felix Gaeta lost a leg, but the bloodstopper kept him alive.

Bloody battle injuries are as old as the first caveman throwing a rock at the second caveman. The natural body process to stop bleeding happens almost immediately. An open wound sends signals into the bloodstream ordering small, tough cells called platelets to converge at the site of the damage. These platelets, also known as thrombocytes, really have no other job but to clump together and swell themselves into spheres to clog up open wounds. An open wound also calls out a net of fibrous material known as fibrinogen, which also migrates to the wound and binds the platelets together into a solid blood clot, preventing the body from losing any more blood.

When the wound is too severe to wait for the body's natural process to take hold, however, it is sometimes possible to help the clotting process along. Starbuck's bloodstopper is almost certainly a sandy granular powder called zeolite, an aluminosilicate mineral with many variants. Zeolites have a unique property that makes them useful in the age of nanotechnology: the pores in a zeolite crystal are very, very tiny and very, very regular. If it becomes necessary to separate individual molecules of one size from molecules of other sizes, zeolite may be used as a natural nano-sized filter that only allows the right size molecules to pass.

A calcium-loaded form of zeolite has pores small enough to let water pass through while leaving other blood material behind. By removing water from the immediate location of the wound, blood-stopper concentrates the platelets and fibrinogen, allowing them to knit together into a clot more easily. The unfortunate side effect is that when calcium zeolite absorbs water, it gives off tremendous amounts of heat. We most likely hear Gaeta scream not from his wound, but from the burning side effect of treating his wound.*

*The man just has no luck at all.

PART TWO

THE PHYSICS OF *BATTLESTAR GALACTICA*

CHAPTER 9

Energy Matters

"Energy," to a scientist, is the ability to do work. "Work," to a scientist, is the application of a force to move a mass some distance. These definitions are pretty far from our casual thinking of "energy" and "work," and in order to understand the concept of energy, we must first understand the concept of work; to understand the concept of work, we must understand the concept of mass.

It has been said that a physics student learns everything three times, and there's no truer example of this than the concept of mass. In its most basic sense, mass can be viewed as the amount of matter an object contains (a count, if you will, of the total number of electrons, neutrons, and protons in an object).

Mass is commonly measured in kilograms. Originally a kilogram was defined as the mass of water in a volume of one liter at 4 degrees centigrade, where water is at its densest. Now it is precisely defined by a block

of metal—a 90 percent platinum, 10 percent iridium alloy—in a climate-controlled vault of the International Bureau of Weights and Measures[1] in the town of Sèvres on the outskirts of Paris. Official copies of the official kilograms are made available by the bureau to other nations to serve as national standards.

Alternately, an object's mass can be viewed as its tendency to resist any change in motion. From experience we know that the more mass an object possesses, the more difficult it is to move. Anybody who has pushed both a shopping cart and a stalled automobile has first-hand knowledge of this. This second definition is often referred to as an object's inertial mass.

An object's *mass*, however, should not be confused with its *weight*. Here on Earth, in everyday conversation with nonscientists, we do not generally differentiate between mass (how much material an object contains) and weight (how the material in an object is affected by a gravitational field). Because the vast majority of us live our lives in Earth's gravitational field, we commonly say that an object "weighs one kilogram," when technically we should say that the object "is being accelerated by Earth's gravitational field by the same amount as a one-kilogram mass is accelerated by Earth's gravitational field." To be perfectly correct, one *could* also say that the object "weighs 9.81 Newtons." No one talks that way because . . . well, try talking that way in any school in the United States, and see how long it takes you to get beaten up.

While the distinction between mass and weight doesn't mean much to the average person, to scientists and engineers, and most especially astronauts, the two concepts are as different as Cylons and Colonials. *Your* weight, right now, is determined by the amount of matter in you and by Earth's gravitational field. Anywhere else in the universe, your weight would be different. The strength of Earth's gravitational field is, in turn, dictated by how much mass Earth itself has.

As an example, take the Apollo astronauts who walked on the Moon. The mass of their spacesuits and backpacks was roughly 80 kilograms. An astronaut whose mass was 80 kilograms would, when suited up,

Admiral Helena Cain, in *Razor.*

Major Kendra Shaw, in *Razor.*

be carrying the equivalent of himself around on his back! Some of the moonwalks lasted seven hours—can you imagine carrying another person on your back for seven hours?

Well, neither could the Apollo astronauts—they didn't have to. Because the Moon is both smaller in diameter and less dense than Earth, it has much less mass than our planet. This translates to much less gravitational attraction at its surface, which means that objects on the Moon weigh one-sixth what they weigh on Earth. This is important: the astronauts on the Moon still had 160 kilograms of *mass*; that didn't change, but the *weight* (the way that mass was affected by a gravitational field) was only about 27kg—a small fraction of what it was on Earth.

There's also a third definition of mass, used by many disciplines within physics: an object's mass is defined solely as its energy equivalent. Before we discuss the interrelation between mass and energy, though, let's return to the concept of work.

In classical physics, *work* is the amount of energy transferred along a distance; for example, moving an object from point A to point B. Energy can be divided into two types: potential and kinetic. Kinetic energy is the energy a body possesses because of its *movement*, the energy that is actually used to do work. Whether it is large in scale, like a planet orbiting a star; medium in scale, like a car rolling down a hill; or minute in scale, like molecules in an object moving more rapidly as they get heated, kinetic energy is nothing but mass in motion. Any moving object has kinetic energy; an object at rest has zero kinetic energy. A released spring, an energized motor, and a yo-yo on its way down its cord are all bursting with kinetic energy. Because of this, kinetic energy can be easily transferred from one body to another. If you've ever touched a hot stove, punched someone in the nose, played billiards, or gotten hit by a train, you know how easy it is.

Potential energy is the energy that an object possesses because of its *condition*. It is energy that is not being put to work now, but is being held, either naturally or by design, to be used later; it is the "potential to have kinetic energy." A diver waiting to leap off the platform, an unburned log, a tightly wound spring, a glow-in-the-dark sticker in sunlight, and a yo-yo held in the downward-facing palm of your hand are all reservoirs of potential energy.

There are many different types of potential energy, and they generally can't be converted to each other, at least not readily and not directly. In our examples just given, the potential energy inside an unburned log is chemical energy. When both oxygen and a source of heat are applied to the log, the log catches fire and burns—giving off enormous amounts of thermal energy and undergoing a chemical transformation in the process. When the diver leaps off the platform, gravitational potential energy accelerates the diver down into the water. Similarly, when a yo-yo is let go, gravitational energy pulls it down to the end of its string (and rotational inertia, or angular momentum, brings it back up). The glow-in-the-dark sticker, when placed in the dark, gives off photons of light it had stored in metastable quantum states when it was in the sunlight.

Kinetic energy can be converted to potential energy and vice versa. In all the cases we discussed, the total amount of energy, the sum of an object's kinetic and potential energies, is a constant. An excellent example of this can be seen in the episode "Exodus, Part II." In order to rescue thirty-nine thousand humans from Cylon-occupied New Caprica, Admiral Adama attempts a daring move: to get underneath the Cylon defenses he orders an FTL jump directly into the planet's atmosphere, beneath the orbiting Cylon Baseships. The jump is executed, and *Galactica* materializes in the sky above New Caprica. Initially at rest with respect to the planet's surface, for a brief moment the ship has no kinetic energy but literally a whole battlestar-load of potential energy. *Galactica* falls. The gravitational potential energy of the Battlestar relative to New Caprica is converted more and more into kinetic energy as *Galactica* accelerates toward the ground. This kinetic energy is further converted into heat energy, as atmospheric friction heats *Galactica*'s underside to the point that its surface vaporizes and ablates. Adama gives the command to launch Vipers. Just before slamming into the ground,

Galactica jumps away, having completed her mission, while her Vipers proceed to apply various forms of kinetic energy (bullets, bombs, and rockets, for example) to the Cylons! What really would have happened to the launching Vipers when they hit the shock wave and plasma enveloping *Galactica*? Who cares when the scene has such a high coolness factor? Repeat the First Law of *The Science of Battlestar Galactica* to yourself: "It's just a show, I should really just relax."

CHAPTER 10

$E = mc^2$

When Albert Einstein derived $E = mc^2$ over a hundred years ago, he told us that at some level, energy and matter are *equivalent*. Just as kinetic energy and potential energy can be converted, one into the other, the most famous equation in all of physics tells us that energy can also be converted into matter, and vice versa. It's the viceversa we'll be discussing here.

In the equation $E = mc^2$, m stands for mass, the amount of material in an object, which we discussed in the previous chapter. The variable c stands for the speed of light, officially measured as 299,792,458 meters per second (it's usually approximated as 3.0×10^8 m/s for all but the most exacting computations). The meter is the basic unit of length in the metric system, and originally was defined by the French Academy of Sciences as 1/10-millionth of the distance from the North Pole to the Equator, when measured on a line through the Paris Observatory. The meter was designed to assist in the derivation of other basic units of the metric system. For example, in the

William Adama.

Laura Roslin.

WHAT IS THE MASS OF BATTLESTAR *GALACTICA*?

Galactica, like all objects, has mass. It spends most of its time in space, far from many gravitational fields, so it generally does not have much weight, but it has mass. Boy, oh boy, does it have mass! Let's estimate how much.

According to the visual effects artist who designed *Galactica*, a Jupiter-class battlestar is 1,371 meters (4,500 feet) in length. It is 538 meters (1,760 feet) wide when the flight pods are extended and 156 meters (543 feet) in height. Let's assume that *Galactica*'s width decreases by 138 meters (452 feet) when the pods are retracted, and call it an even 400 meters wide. A box 1,371 by 156 by 400 meters encloses a volume of 85,550,400 cubic meters. Since *Galactica* isn't a perfect

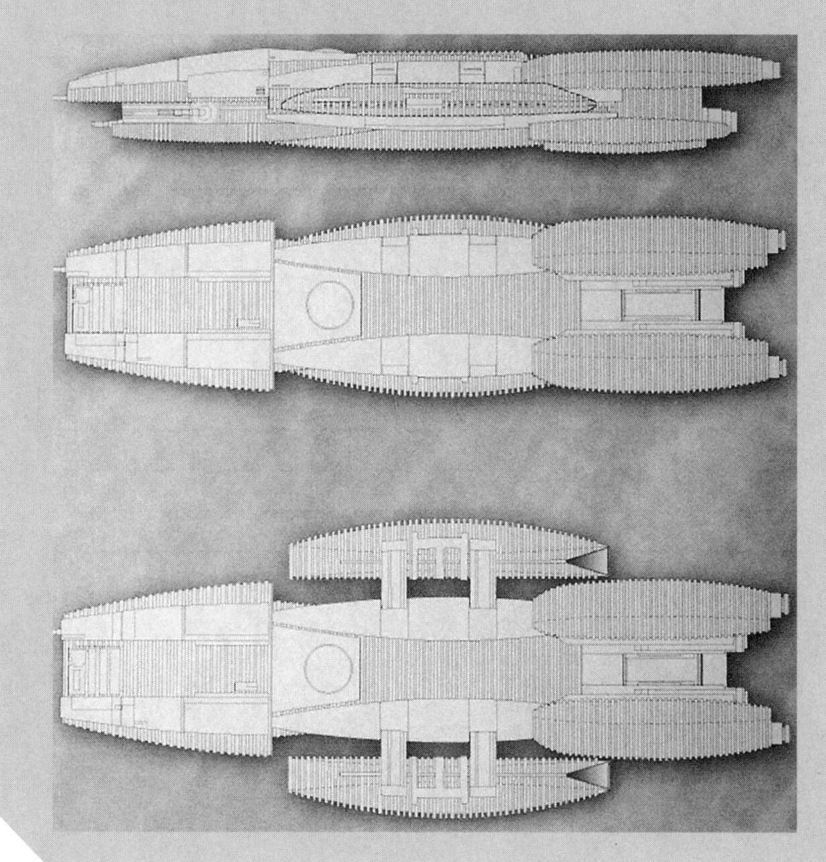

A schematic diagram of Battlestar *Galactica*.

»>

rectangle, let's take 15 percent off that figure, and say that the volume of *Galactica* is 72,717,840 m^3.

The table below shows the mass of a *Galactica-sized* volume filled with various materials.

If *Galactica* were made completely of:	At a kg/m³ density of:	It would have a mass in kilograms of:
Sawdust	210	23,651,271,000
Manure	400	45,050,040,000
Powdered Milk	449	50,568,669,900
Soap	481	54,172,673,100
Apples	641	72,192,689,100
Plaster	849	95,618,709,900
Butter	865	97,420,711,500
Rubber	945	106,430,719,500
Water	1,000	112,625,100,000
Cement, Portland	1,506	169,613,400,600
Magnesium	1,738	195,742,423,800
Beryllium	1,840	207,230,184,000
Ivory	1,842	207,455,434,200
Brick	1,944	218,943,194,400
Aluminum	2,560	288,320,256,000
Titanium	4,500	506,812,950,000
Steel, Stainless	7,480	842,435,748,000
Iron	7,850	884,107,035,000
Brass	8,430	949,429,593,000
Bronze	8,780	988,848,378,000
Nickel	8,800	991,100,880,000
Lead	11,340	1,277,168,634,000
Depleted Uranium	18,900	2,128,614,390,000
Gold	19,320	2,175,916,932,000
Plutonium	19,800	2,229,976,980,000
Platinum	21,400	2,410,177,140,000

»»

In estimating the mass of *Galactica*, we start by estimating the mass of her armored exterior hull. The key questions, then, center around (1) the thickness of the hull and (2) the composition of the hull. Obviously, those two elements are related.

Almost immediately we come across some problems. In the Miniseries, *Galactica* took a direct hit from a Cylon tactical nuclear weapon, but it didn't vaporize *Galactica*. This tells us two things: the Cylon nuke wasn't very big, and *Galactica*'s hull is very strong. More precisely, *Galactica*'s hull is somehow able to deal with the enormous energies given off by even a small nuke. When Starbuck eyeballs the result, she reports that the forward section of the port flight pod has sustained heavy damage. She (and we) see lots of relatively small hull breaches, lots of escaping atmosphere and smoke. We do not, however, see a crater. The Cylon nuke did not gouge out any sort of hemispherical opening in the hull. How can that happen?

During its heyday in the 1950s, nuclear weapons testing provided scientists with data

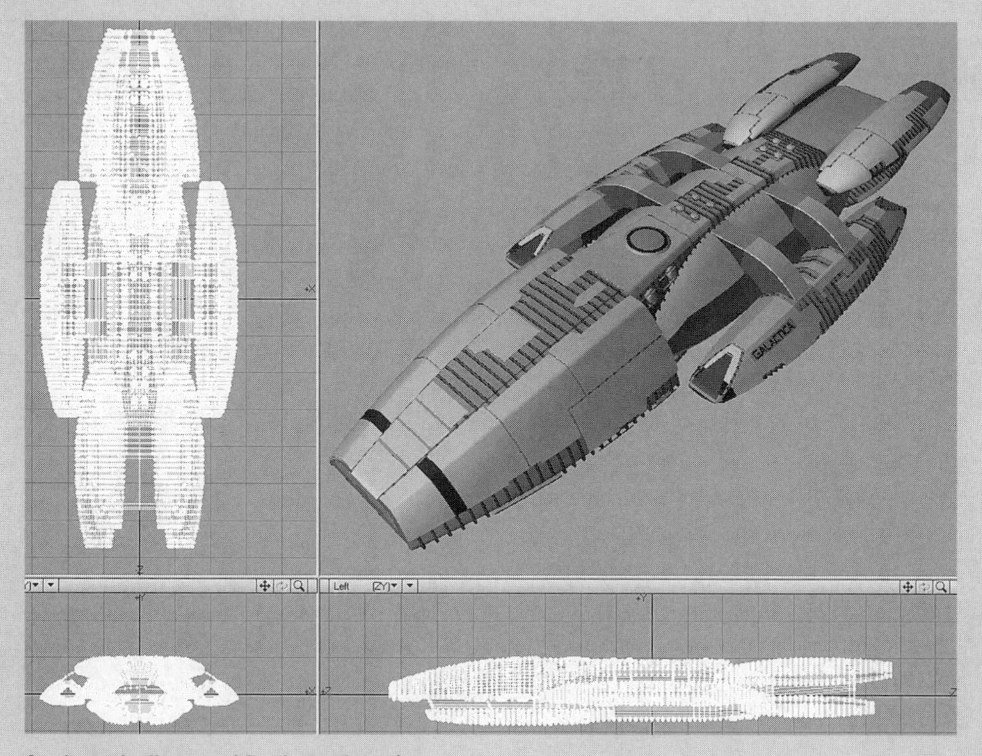

A schematic diagram of Battlestar *Galactica*.

»»

not just on the nuclear explosive itself, but also on its effect on the surrounding environment.* From this data, scientists have built up a set of formulae that we can use to estimate the effect a tactical nuke might have on *Galactica*. From the damage estimates in this episode and in the episode "Water," it seems that the nuke detonated at the forward end of the flight pod, near the water bulkheads in the ship's "alligator head." Since the flight pod is 600 meters long and the damage seems confined to the forward quarter, we can estimate that the area of destruction was caused by a 10-kiloton nuke, detonated 3 meters from the hull.

Hull material	Crater depth (meters)
Aluminum	12.6
Carbon-titanium	11.9
Uranium	7.1
Titanium	8.3
Tungsten	4.7
Iron	14.9

The figures are even worse for a nuclear detonation at zero meters from the hull. For those cases, the hull crater would be on the order of 100 meters deep and would essentially have cut *Galactica* in two. Bad enough as these figures are, even if *Galactica* were made of pure tungsten, it would need to have an outer hull approximately 5 meters thick throughout the ship. Let us also assume that decks and bulkheads fill 10 percent of the interior. They have to be strong, but not as tough as the exterior, so we assume that they are composed of structural steel, which also happens to be lighter than tungsten. We can now estimate the mass of *Galactica* under those circumstances:

- Outer measurements: 1,371 by 156 by 400 meters = 85,550,400 m^3
- Inner measurements: 1,366 by 151 by 395 meters = 81,475,070 m^3
- Estimated volume of tungsten hull: 4,075,3300 m^3
- Estimated mass of *Galactica*'s hull: 78,845,010,000 kilograms
- Minus 15 percent (because *Galactica* is not perfectly rectangular): 66,682,587,100 kg
- Estimated volume of decks and bulkheads: 6,925,400 m^3

* Watch any documentary on nuclear weapons and you'll see the same film clips over and over—wood-frame and brick houses, built specifically for the test and sitting like pariahs in the middle of the Nevada desert, implode and catch fire when the bomb goes off. Fake pine forests, planted in cement like Mafia victims, sway and break in the blast wave.

»»

- Mass of decks and bulkheads: 54,433,494,300 kg
- Estimated mass of *Galactica*: 121,116,081,400 kg

One hundred and twenty-one *billion* kilograms? That's the mass of a small asteroid! We haven't even counted the additional mass of *Galactica*'s interdecks, bulkheads, engines, equipment, fuel, water, and people, but no matter; they're dwarfed by the sheer mass of the tungsten hull and structural members. So let us assume we have a working mass for *Galactica*: 121 billion kg.❮

original metric system definition, a kilogram was defined as the mass of water that could fit in a box one-tenth of a meter on each side. Later, the meter was defined as the distance between two marks on a platinum-iridium rod next to the six one-kilogram cylinders stored in the BIPM vaults. Since 1983, however, the meter has been defined, in a somewhat circular fashion, as the distance traveled by light in a vacuum in 1/299,792,458ths of a second.*

The variable *E* stands for energy. In the International System of Units, energy is measured in *joules*. A joule is the energy required to move a one-kilogram mass vertically one meter in a gravitational field of one meter per second. Gravity of that strength would be about one-tenth Earth gravity, or slightly less than lunar gravity. Another way to put it: it's the energy required to lift an apple one meter straight up from Earth's surface, or the energy realized when an apple drops one meter to Earth's surface. Since Earth's gravity is 9.81 meters per second per second at sea level, one kilogram at one meter above Earth's surface has a potential energy of 9.81 joules.

The following table takes a look at various masses and the energy released when they are converted entirely to energy.

---

* And exactly what is a second, you may ask? A second is defined as the duration of 9,192,631,770 periods of the radiation corresponding to the transition between the two hyperfine levels of the ground state of the cesium 133 atom.

Item	Mass (kg)	Energy (joules)	Equivalent
Macintosh apple	0.1	8.987×10^{15}	~2 megatons TNT
Bar of soap	0.13	1.146×10^{16}	2.75 megatons
Small dog	2.5	2.2469×10^{17}	53 megatons—roughly equivalent to the largest nuclear device ever exploded
Apple Macintosh 128 k	7.5	6.7407×10^{17}	161 megatons
Sailboat anchor	15	1.348×10^{18}	Total U.S. electricity use per year
Adult horse	520	4.6735×10^{19}	U.S. yearly oil consumption
Olympic keelboat	675	6.0486×10^{19}	World electricity use
Toyota Sienna XLE	2,000	1.7975×10^{20}	World yearly oil consumption
M115 155 mm howitzer	5,600	5.0330×10^{20}	Total yearly global energy use
Unloaded 747 aircraft	175,000	1.5728×10^{22}	Total energy of sunlight falling on Earth each day
Galactica	1.21×10^{11}	1.0874×10^{28}	Energy output of the sun for 27 seconds

The speed of light is a very large number, and that number squared is a huge number. When these quantities are multiplied together, it's easy to see that a conversion of even a tiny amount of matter will yield enormous amounts of energy.*

* By squaring the speed of light, we are multiplying the mass by approximately 300 million times 300 million. That's a *lot* of energy. $c^2 = 8.99 \times 10^{16}\ m^2/s^2$.

CHAPTER 11

Special Relativity

Great leaps in science are rarely, if ever, done by consensus or by large groups. While nearly every scientist, working alone or in a team, contributes to the sum total of human knowledge, and while there is the occasional researcher who simply happens to be in the right place at the right time, often quantum jumps in our understanding of nature are made by a lone researcher who simply understands an important aspect of nature far better than everybody else . . . and can *prove* it. Albert Einstein was one such visionary. In 1905 he published an article entitled "On the Electrodynamics of Moving Bodies"[1] that turned the scientific world on its head. It was in that article that Einstein first published his Special Theory of Relativity.

> **So the universe is not quite as you thought it was. You'd better rearrange your beliefs, then. Because you certainly can't rearrange the universe.**
> —Isaac Asimov and Robert Silverberg, *Nightfall*

It had only been about three hundred years since science—an organized methodology for understanding and describing nature—had clawed its way out of the superstition of the Middle Ages. Based upon observable phenomena, science relies on such concepts as measurement, objectivity, and reproducibility of results. Special Relativity challenged two of those very pillars of science: measurement and objectivity. In fact, one implication of Special Relativity (SR) is that measurement is inherently *subjective*.

There are two postulates of Special Relativity. The first one states that there is no such thing as an absolute definition of motion in the universe, meaning also that there is no absolute standard of rest, either. Motion (or rest) must, therefore, be specified relative to *something*.

Think of a Viper about to be launched at a basestar. Relative to *Galactica*, in its *reference frame*, the Viper is receding from the ship. To the Cylons, the Viper is inbound and CBDR* in their reference frame.

Although neither *Galactica* nor the Basestar was moving relative to each other, the Viper was moving relative to both of them. In the Viper's reference frame, however, both Galactica and the Basestar were moving and the Viper was stationary. In Special Relativity, all three viewpoints are equally valid.

The second postulate of Special Relativity states that the speed of light in a vacuum is a universal invariant† and is, as far as we know, the speed limit in our universe. As we have seen, photons have zero mass and propagate through a vacuum at 299,792,458 meters per second, or 186,282 miles per second (the speed of light, a value for which we normally use the variable *c*‡)—though they travel more slowly through

*Recall from the episode "Final Cut" that CBDR means "constant bearing/decreasing range"—a collision course.

†The term "speed of light" is somewhat parochial: a more correct term is the "speed of electromagnetic radiation" or the "speed of photons." See chapter 13.

‡*c* for *celeritas*, Latin for "swiftness." A beam of light could travel around the circumference of Earth almost seven and a half times in one second. Think of that next time you're waiting for an email to arrive.

Starbuck holding the Arrow of Apollo, with Six looking on.

Kara "Starbuck" Thrace.

other transparent media like air, glass, and water.* No object that has mass (even the tiniest subatomic particle) or carries information can travel faster than the speed of light—in fact, no object with mass can even travel *at* the speed of light. We'll explore how this rule may be broken, or at least severely bent, in chapter 22.

The simple fact that light has a finite speed has profound implications. Most importantly, it implies that space and time are inexorably linked. One fascinating implication of *this* is that when you look out into space, you also look back into time: the farther out you look, the farther back you look.

For example, our nearest celestial neighbor is the Moon, which is roughly 384,000 kilometers away. It takes reflected sunlight about 1.3 seconds to reach us after bouncing off the Moon. This means that when you see the Moon, you aren't seeing it as it is *now*, you are seeing it as it was 1.3 seconds *ago*. It takes sunlight nearly eight and a half minutes to reach us; therefore, when you look at the Sun† you do not see it as it is at present, you see it as it was when photons first left its surface eight and a half minutes ago. The brightest star in Earth's night sky is named Sirius,‡ which is in the constellation Canis Major. The light from Sirius takes 8.6 years to reach us, so you're seeing it as it was almost a decade ago. That distance is so enormous that astronomers have had to come up with a new metric, called a light-year.

Contrary to common misunderstanding, a light-year is not a measure of *time*. It is a measure of *distance*—the distance that light travels in one Earth year, roughly 9.5 trillion kilometers. It's kind of numbing to think that Sirius, one of the closer stars, is roughly

---

*In 2001 a team of researchers passed a beam of light through a substance called a Bose-Einstein condensate (BEC) and slowed the measured speed of the beam to a mere 17 meters per second. The individual photons actually still traveled at c, but the atoms within the BEC temporarily captured and reflected the photons so often that they appeared to move through the medium very slowly.

†Don't look directly at the sun.

‡For some reason, vast numbers of people think that the "north star" is the brightest star in our sky. The truth is that Polaris, the current North Star, is barely in the top fifty brightest stars. The brightest is Sirius. Seriously.

81,700,000,000,000 km away—we instead say it is 8.6 light-years away and go on to other things.

When we discuss travel over the vast distances between star systems, we think in terms of traveling through four-dimensional spacetime: three dimensions of space and one dimension of time.*

An observer with a keen eye would have noticed how this was used in the episode "Lay Down Your Burdens, Part II." Newly elected president Baltar signed the executive order to settle on the planet New Caprica. Although New Caprica was something less than hospitable, because it orbited a star embedded within a nebula, many believed that it would be nearly impossible for the Cylons to detect. At around this time Gina Inviere—a Cylon Six—detonated a nuclear warhead aboard the luxury liner *Cloud Nine*, taking two other ships and 4,400 souls in the process.

Although life was hard for the following year, the humans were hanging onto the belief that they had gone unnoticed by the Cylons. One ordinary day, without warning, an entire Cylon armada jumped into orbit, beginning the occupation of New Caprica. When the Cylons arrived on *Colonial One*, President Baltar asked their welcoming committee the obvious question: "How did you find us?"

A Five responds: "Quite by accident, actually. We were over a light-year away from here when we detected the radiation signature of a nuclear detonation."

It took the radiation from that explosion just over a year to reach them. As soon as the Cylons detected the gamma radiation pulse from the nuclear explosion, they determined the direction from which the pulse came, rapidly assembled a fleet, and immediately jumped to New Caprica. That part took only a few hours or days; it had already taken the light a year to reach them.

The second tenet of Special Relativity—the speed of light in a vacuum is invariant—means that it is a universal constant. Two observers will always measure the exact same value for c no matter what their relative motion. While this may not seem particularly

*There is a fifth dimension, beyond that which is known to man. But that's for another book entirely.

counterintuitive initially, it has some mind-boggling implications. If the speed of light is constant, that means that some other values we think of as constants—an object's physical length, its mass, even the rate at which time passes—are variable. This is best clarified by an example.

Suppose Centurions have boarded *Galactica*. There is a firefight in the hangar bay with the Colonial Marines at the front of the bay and the Cylons at the rear. Both sides fire bullets that travel at 1,000 meters per second. You watch the firefight from the comparative safety of *Rising Star* as *Galactica* drifts slowly past at a relative speed of 10 meters per second. If you could measure the speed of the bullets the Cylons were shooting at the Marines relative to you, you would measure the bullets' muzzle velocity (1,000 meters per second) plus the forward motion of *Galactica* (10 meters per second) for a combined velocity of 1,010 meters per second. If you could measure the speed of the bullets that the Marines fire at the Cylons—from the front of the hangar bay to the rear—you would measure their muzzle velocity minus *Galactica*'s forward velocity, or 990 meters per second. Nothing unexpected so far . . .

Now switch reference frames. Specifically, go back to the 1978 *Battlestar Galactica*, when both sides had guns that shot beams of light and went "Pew! Pew!" In this world, suppose Centurions have boarded *Galactica*. There is a firefight in the hangar bay with the Colonial Warriors at the front of the bay and the Cylons at the rear. You watch the firefight from the comparative safety of the *Rising Star* as *Galactica* drifts slowly past at a relative speed of 10 meters per second. Recall that in the original series, both the Colonials and the Cylons had projected energy weapons. Since the energy that comes from these types of weapons is a form of electromagnetic radiation, it naturally travels at the speed of light. If you were to measure the speed of the LASER pulses the Colonial Warriors were firing at the Cylons, you would measure c.* If you measured the speed of the pulses the Cylons shot at the Warriors, you would also measure c. No matter how fast you

*LASER stands for Light Amplification by Stimulated Emission of Radiation.

were traveling—even if you were Apollo in his Viper zooming past the firefight—the speed of the laser blasts would be constant.

However, if the speed of light is a constant, irrespective of the relative motion of the source of the light and the receiver, then something else has to be flexible in order to make that so.

Time Dilation

Over the years many a controversial* scientist, attempting to have his or her views embraced by mainstream science, has fallen back on the apocryphal argument "They laughed at Einstein at first, too!" But there's no historical reference of anybody laughing at Einstein.† While the implications arising from much of Einstein's work were groundbreaking, counterintuitive, and changed the way we look at the very fabric of our existence, they were mathematically sound—and to a scientist, a mathematically rigorous argument is a convincing one. Einstein's work spoke for itself.

If the mere mathematics of Einstein's formulation of Special Relativity failed to prove compelling, several of the implications—like relativistic time dilation—have been verified experimentally. One outcome of SR is that time moves more slowly for you if you are moving relative to a reference that you have defined as "stationary": a clock in motion moves more slowly than a clock standing still. As your speed increases to a large fraction of the speed of light, you are said to be traveling relativistically. When the passage of time is slower for a high-speed object, this is called *relativistic time dilation*. Experimental validation of time dilation was provided by rain: a rain of muons.

Muons are subatomic particles that have a negative charge and are like short-lived "cousins" of electrons: on average they decay into less massive subatomic particles in slightly over 2.2 microseconds.

---

*Read: crackpot.

†Except maybe at his hair. And his absent-mindedness. And his lack of socks. And the way he would put a whole egg in his soup, boil them both for ten minutes, and have a meal of hot soup and a hard-boiled egg. But nobody laughed at his work.

Starbuck in her Viper.

They are created when high-energy cosmic radiation from space interacts with Earth's upper atmosphere. With such a short life span, even traveling at 0.99*c* most muons would travel an average distance of 650 meters before decaying. Few, if any, would ever reach the ground. Yet experiments have found that numerous muons, in fact, impact the surface of Earth *because* they travel at nearly the speed of light. Their lifetimes are extended due to time dilation.

Suppose Starbuck and Apollo rocket out of the launch tubes in their Vipers, bank in opposite directions, put a fair distance between them, then turn to face each other and come to a relative stop. Starbuck fires her engines to get up to her Viper's top speed and coasts past Apollo. Once Starbuck is coasting—once her Viper has attained a state of uniform, nonaccelerated motion—we can ask ourselves, "Who is moving?" In Apollo's frame, Starbuck is moving toward him. In Starbuck's reference frame, it is Apollo who is approaching. Because of time dilation, as they pass by each other, they will disagree about the rate at which time passes. The "other" observer's clock will seem to be slower: If Starbuck could see the clock on Apollo's Viper, she would see that it was moving slower than hers. If Apollo could observe Starbuck's clock, he would see her clock running slower than his! Seems impossible? Welcome to the wonderful world of Special Relativity.

This effect is more pronounced the faster the relative motion. Given two observers, like Apollo and Starbuck in our example, the equation for time dilation is:

$$\Delta t_{moving} = \frac{\Delta t_{rest}}{\sqrt{1 - \frac{v^2}{c^2}}}.$$

This equation shows that the effect of time dilation gets markedly more pronounced as the relative motion of the two objects, in our example Vipers, approaches the speed of light. We see this also in the table to the right. If, for example, Kara's Viper shot past Lee's at 75 percent of the speed of light, and if she could view the clock on his Viper, she would perceive his clock moving at 66 percent of its normal speed. Lee, looking at Kara's ship, would observe the same thing—Kara's clock is running at 66 percent of its normal speed.

Sam Anders referred to time dilation explicitly in the episode "No Exit" when discussing how the Final Five could travel from Earth to the Twelve Colonies while aging minimally: "Time dilation. We hadn't developed Jump drives, so we traveled at relativistic, but subluminal, speed."

We now understand how the Final Five can be over three thousand years old, but even the oldest of them doesn't look a day over sixty-five!

Another allusion to time dilation appeared in an early draft of the episode "He That Believeth in Me." When Kara returns from the dead, she has some obvious explaining to do in order to convince Admiral

Time dilation as a function of relative speed

Velocity as a fraction of *c*	Δt_{rest}	Δt_{moving}
0.1	10 sec	10.1 sec
0.5	10 sec	11.5 sec
0.75	10 sec	15.1 sec
0.9	10 sec	22.9 sec
0.99	10 sec	70.9 sec
0.999	10 sec	223.7 sec

Adama, among others, that she is not a Cylon—particularly when her Viper appeared "as if it were off the showroom floor." Although two months had elapsed for the crew of *Galactica*, only six hours had elapsed on her Viper's onboard clock. Kara's explanation: "Look, I flunked the temporal relativity quiz in space-flight physics. I don't understand the time disparity either."

Although this was a fun bit of exposition, it was also unnecessary to move the plot forward, so it was deleted in later drafts. Still, Kara's line, as well as Anders's line earlier, clearly shows that *Battlestar Galactica* was one of the rare science fiction TV series to make dramatic use of the bizarre implications of Special Relativity.

Lorentz Contraction

Measure the length of an object, like a spacecraft, at rest and you'll come up with its rest length. Measure the length of an object, like a spacecraft, at relativistic speeds and you would measure a shorter length. The faster the object moved past you, the shorter its length would be. This effect is called the Lorentz Contraction or, more correctly, the Lorentz-Fitzgerald Contraction. If it seems unusual that time can travel at different rates for two objects with a high relative motion, it probably seems even more bizarre that spatial extent—an object's length—is similarly variable. Then, again, given that space and time are intimately linked, perhaps this shouldn't be surprising.

Similar to time dilation, the amount of physical contraction gets much more pronounced as the moving object approaches the speed of light. The equation for the Lorentz Contraction is:

$$L_{\text{moving}} = L_{\text{rest}}\sqrt{1-\frac{v^2}{c^2}}\,.$$

Again, this is best elucidated by example. The Final Five Cylons are escaping the nuclear fires of Dead Earth and heading to the Twelve Colonies. Since their ship was not jump-capable, but was capable of traveling at relativistic speeds, they have time on their hands. A lot of time. Since they are all scientists, there are many discussions and debates along the way. Along the way, Galen and Saul get into a spirited debate about how their speed is affecting the passage of time and the length of their ship. So when they stop briefly at the Algae Planet to stock up on a few things (and build a temple), they decide to do a quick experiment. Tory will accelerate the ship up to a predetermined speed. At a set time (and knowing this group, we're assuming that there was another long debate on how this would be accomplished), Tyrol would measure the length of their ship, whose rest length is 100 m, while onboard. Tigh would be stationed in a shuttle and would measure the length of the ship as it passed by. We'll assume that the ship's top speed was 0.999*c*. After six trials at six different speeds, Tyrol and Tigh compared their measurements.

Lorentz contraction as a function of relative speed

Velocity as a fraction of *c*	L_{rest} (Tyrol)	L_{moving} (Tigh)
0.1	100 m	99 m
0.5	100 m	87 m
0.75	100 m	66 m
0.9	100 m	43 m
0.99	100 m	14 m
0.999	100 m	4.5 m

Tigh sees the ship successively shorter the higher the relative speed. Recall that motion is meaningful only relative to a given reference frame. Anders, who was doing an experiment of his own, also noted that Tigh's ship was shortened by the exact same fraction.

Let's return to our example with Starbuck and Apollo. Assume that when they passed at 0.75*c*, not only did they observe each other's clocks, but they measured the length of each other's Vipers. Lee says that Kara's Viper was relativistically shortened, and Kara says that Lee's Viper was. From each other's point of view, both Vipers were compressed to 66 percent of their length at rest. (The same percentage by which their clocks slowed down, notice.)

The effects of Special Relativity seem irrational simply because humans don't travel at relativistic speeds yet. These effects may not be intuitive, but they do describe the way the universe works on a large scale. In fact, if the implications of Special Relativity seem like something from *The Twilight Zone*, things are about to get weirder as we turn our attention to gravity and General Relativity.

CHAPTER 12

General Relativity and Real Gravity (or the Lack Thereof)

Nearly all spacecraft on TV or in the movies have some form of artificial gravity. This type of artificial gravity, built not around physics but economics, was invented around the turn of the last century by a French stage magician named Georges Méliès. Méliès left the theater to become a trailblazer in the then-new medium of motion pictures, and in 1902 he produced the science fiction classic *Le Voyage dans la Lune*. The film portrayed a Jules Vernean voyage by a group of crazy wizard-scientists (aided by a bevy of pretty girls) launched by rocket cannon from Earth to the Moon.

The excessive G forces involved with launching people from a cannon—which should have flattened them into a puree—apparently didn't affect these intrepid *voyageurs*. When Méliès's astronauts land on the Moon, they are clearly not jumping around under the diminished influence of lunar gravity. Like many subsequent movie producers, Méliès either didn't know, or didn't care, about the changes in the magnitude of gravity that take place when one leaves the surface of Earth. And in nearly every science fiction film that followed, from the *Flash Gordon* serials to *Duck Dodgers in the 24½th Century*, cosmic travelers kept their feet firmly planted on the deck of their spaceships. If anyone bothered to ask, "artificial gravity" became the easy way out for movie producers who didn't want to spend hundreds of thousands of dollars rigging wires, using other complicated special effects, or renting a special NASA low-gravity-simulator aircraft, nicknamed the "vomit comet," to give the illusion that their actors were weightless. Simply turn on the "artificial gravity" and presto—actors could walk around a soundstage in Hollywood and not have to explain to the audience why they weren't experiencing the near-zero-gravity environment of outer space.

Why do real-life astronauts, working on the flight deck of the space shuttle or within the modules of the International Space Station, float? What is weightlessness anyway? For that matter, what is gravity? For that answer, we turn to both Sir Isaac Newton and Albert Einstein.

In 1687 Sir Isaac Newton published his best-known work, his *Philosophiae Naturalis Principia Mathematica* (Mathematical Principles of Natural Philosophy), often known simply as *The Principia*. It was in the *Principia* that Newton first published his Law of Universal Gravitation, which dictated the gravitational force between any two objects. The Law of Universal Gravitation is one of the most famous equations in physics:

$$F = G\frac{m_1 m_2}{r^2}.$$

In this equation, F represents the gravitational force between the two objects, G is the universal gravitational constant,[1] m_1 and m_2 are the masses of each object, and r is the distance between the two objects. Simply stated, any two objects in the universe that possess mass attract each other via the force of gravity. Even the book in your hands exerts a small gravitational pull. The more massive the objects are, the greater their mutual gravitational attraction.

The force of gravity between two objects decreases dramatically as the distance between them increases, in what we call an inverse-square relationship. For example, Saturn is approximately twice as

Chief Galen Tyrol and Sharon "Boomer" Valerii.

Chief Galen Tyrol and Sharon "Boomer" Valerii.

far from the Sun as Jupiter. Since the force of gravity decreases as $1/r^2$, Saturn feels about one-quarter the gravitational attraction from the Sun as does Jupiter. Neptune, thirty times farther from the Sun than Earth, feels 1/900 the gravitational pull of the Sun that Earth does. An important thing to note is that Newton's Law of Universal Gravitation is not time-dependent. In other words, if Newton's equations were correct, were the star Sirius to disappear right now, you would instantly stop feeling its gravitational attraction

Newton's Law of Universal Gravitation has not been entirely abandoned, and still has wide application in physics and engineering today (like planning the trajectories of spacecraft to other planets). In fact, it will come in very useful in our subsequent discussion of artificial gravity later in this chapter. In the years following Newton's *Principia*, however, it became obvious that his law was not a complete description of gravity.

When Special Relativity arrived on the scene in 1905, there were apparent contradictions to Newton's Universal Law of Gravitation. So in 1915, and again in 1916, Einstein published two papers expounding upon the General Theory of Relativity (GR) to reconcile the two theories. The fundamental principle behind General Relativity initially seems to be so straightforward as to be child's play, but anything studied in sufficient detail becomes amazingly complex, and the implications of GR are profound.

According to Einstein's first postulate of General Relativity, there is no difference between gravity and a uniformly accelerated reference frame. That's it. This is easily elucidated by example. Captain Louanne "Kat" Katraine sits in her Viper, preparing to launch. Over the wireless she hears the voice of the catapult officer as he finishes his checklist, "Nav con green. Thrust steady and positive. Mag cat engaged. Good hunting, Kat." The magnetic catapult engages and rockets her Viper down the launch tube, the forward acceleration slamming her back against the ejection seat. Now let's say that on a subsequent launch Kat's crew chief covered her Viper canopy with duct tape so she can see nothing but the inside of her cockpit. Let's also assume that, upon launch, she is typically thrust back into her seat with a force equivalent to the force of gravity, or 1 G. With no external reference,

Kat would not be able to tell if she had been shot down the launch tube, or if her maintenance crew had pulled a prank and simply tipped the Viper up onto its engines and she was forced into the seat by gravity. A person with no outside frame of reference has absolutely no way of distinguishing one from the other,* and the universe sees the two situations as synonymous. The force you feel from gravity is the same as the force of uniform acceleration.

Why was this such a ground-breaking concept? Let's examine the implications. Imagine that we have a super-high-tech elevator, one that is capable of accelerating upward very rapidly (see the figure below). We fire a single photon (see chapter 13, "The Wonderful World of Radiation") into the elevator at the very instant the elevator shoots upward. It will follow a curved trajectory relative to the elevator. Because the force of gravity is equivalent to uniform acceleration, if the elevator had been sitting on the ground the massless photon we fired into it would still have a curved trajectory, this time due to gravity rather than acceleration.

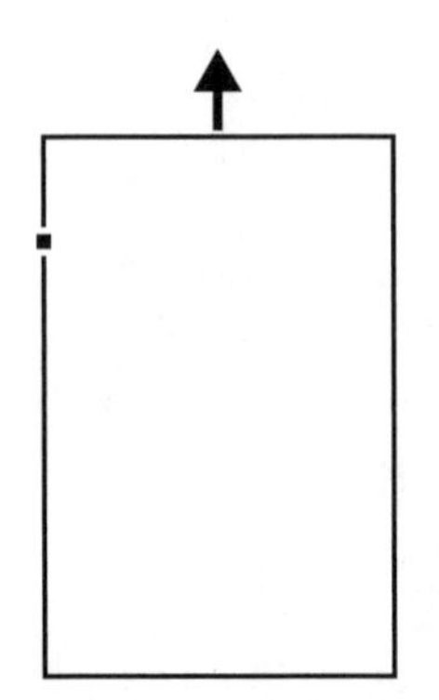

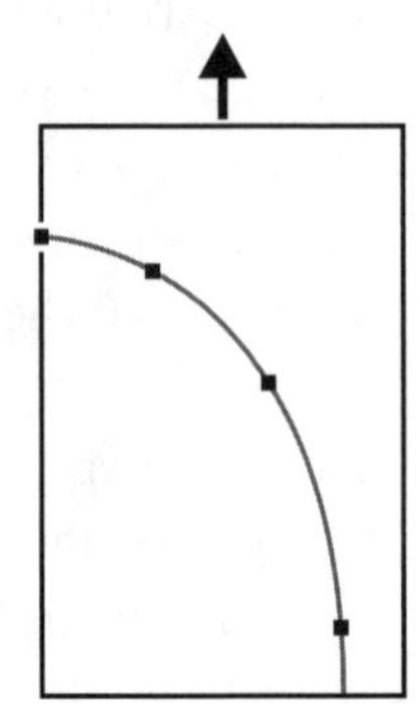

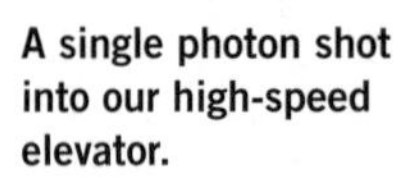

A single photon shot into our high-speed elevator.

According to General Relativity, the force of gravity is not really a force at all, but is instead a warp or curve in the very fabric of space-time. Any object that has mass warps space-time—more mass means more of a warp. Another nearby object in space (and time) will feel that warping of space—as well as creating its own gravity warp—and the two objects will be accelerated toward each other. Therefore, if a large mass like a star warps space-time, then *anything* passing by the star—including light—would be subject to its gravitational influence and have its trajectory altered.

The predicted bending of light, in particular starlight, was experimentally verified in 1919 by Sir Arthur Stanley Eddington. On an island off the west coast of Africa, Eddington photographed a total solar eclipse. Before the eclipse he also photographed the stars that

*We realize that she'd feel the Viper being tipped back, and she'd also feel the vibrations of her Viper on the rails at launch. Work with us here.

should be in the line of sight of the Sun during the eclipse. The apparent positions of the stars near the Sun were shifted by the amount predicted by General Relativity, which revealed that the path of starlight had been gravitationally bent. Similar to the bending of starlight, when NASA spacecraft send data to Earth, they must aim their communications antennae in such a way to account for the curvature of the radio signal due to the gravitational effect from the Sun.

An enlightening way to visualize gravity is to imagine a perfectly taut, perfectly flat trampoline. Place a marble on the edge of the trampoline—it will make a very slight, nearly imperceptible, indentation on the trampoline. You could roll your marble from one side of the trampoline to the other with a mere flick of your thumb. The marble will travel straight and true, and will be slowed, or deflected, only a small amount due to friction with the surface of the trampoline.

Now place a bowling ball in the center of the trampoline. It will cause a sizable indentation. The depth of the indentation depends on the mass of the bowling ball—a 10-pound ball will make a medium sized impression, while a maximum regulation 16-pound ball will make a much deeper "warp" in the trampoline surface. The indentation of the bowling ball in the surface of the trampoline is similar to the way any object with mass distorts the region of space-time around it. The more mass, the greater the distortion.

With the addition of the bowling ball, your marble-shooting game becomes significantly more challenging. Instead of shooting marbles from one side to the other in a straight line, you've now got to deal with this giant frakking indentation in the middle of the trampoline. Any marble rolled across the trampoline no longer travels in a straight line. Instead all marbles curve to

How the warping of space-time by our Sun affects spacecraft communications.

varying degrees toward the indentation created by the bowling ball. Some marbles, launched far away from the bowling ball, are deflected hardly at all. Other marbles, launched with great speed nearer to the ball, are deflected slightly—they make it to the other side, but their paths always curve in the direction of the bowling ball. Marbles traveling at a much slower speed are pulled into the bowling ball completely and never make it to the other side.

Sometimes, with practice or by luck, you can launch a marble at just the right speed and at just the right medium distance from the bowling ball that it travels across the trampoline in ways that would be impossible without the influence of the bowling ball. You can launch a marble in such a way that it curves 90 degrees around the bowling ball. You can even launch a marble so that it travels completely around the bowling ball and returns back to you without bouncing off any other obstacle.

Massive objects (like stars and planets) warp the fabric of space-time like the bowling ball does to the trampoline. More mass means

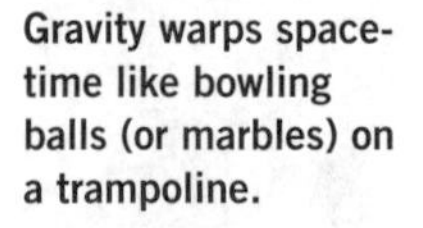

Gravity warps space-time like bowling balls (or marbles) on a trampoline.

a greater warp. Say we have a Raptor traveling in interstellar space far from any massive object. If the Raptor does not fire any thrusters, it will travel in what seems to be a straight line.* That Raptor will find its path deflected as it approaches a star or planet. The Raptor will travel faster and faster and its path will be deflected more and more toward the object's center as it gets closer to the object. If the Raptor is traveling relatively fast to start with, it is only slightly deflected from its path before the encounter. But if the Raptor travels along a slightly different trajectory, either with less velocity or angled more directly at the planet, it will simply be pulled directly toward the center of the planet and the pilot will earn an immortal place in the Colonial Navigation Hall of Shame.

Of course, gravity doesn't only affect planets and spaceships. Once an astronaut's body overcomes its initial confusion and gets adapted to weightlessness, the really bad stuff starts—and it's all because the human body is a lazy sack of bones that won't work at anything it doesn't have to. Once you're in microgravity, your muscles start to come to the realization, "Hey! I don't have to stretch and pull to keep this body upright anymore. I don't have to climb stairs, or perform the careful nonstop push-me-pull-you balancing act of sitting on a chair. I can just relax into a semi-fetal hunch and float. Cool!" More dangerously, your heart says pretty much the same thing: "Woohoo! No more fighting gravity to pump blood to the extremities! I can just take it easy and let the blood and other body fluids pool in the chest!"

Unfortunately, having evolved over billions of years in a gravitational field, the human body becomes confused by the sudden absence of the pull of gravity. Not being smart enough to have worked weightlessness into its repertoire of things that can possibly go wrong, the only thing your body knows when you're floating in space is that there is too much blood, and other body fluids, in your chest. When the same thing happens on Earth, that signals that you are overhydrated. Therefore, when you're in space, your body makes the mistake

* Of course, there is no point in space that is totally isolated from any gravity field, so the Raptor will never travel in a perfectly straight path, but for significant distances the Raptor's trajectory can be a very good approximation of a straight line.

of treating a space symptom like an Earth symptom, and acts accordingly. You pee. A lot.

Your bones, having heard everything the muscles are saying about not straining, decide to take life easy as well. "The muscles aren't pulling as strongly against us," your bones say. "We don't need all this calcium to keep this body upright. Let's get rid of some. Hey urine, a little help here?" Since your body already has gotten the signal to urinate prodigiously, the bones take advantage of this and start to shed calcium and potassium. Your kidneys, not used to all these minerals flowing out in such concentrations, do their best to handle the flow, but they also start to protest. Meanwhile your heart says, "Hang on, I need that potassium to maintain my rhythm—oh no, wait! I'm carrying such a light load, I don't need much potassium. Piss away!" The problem is that if you do suddenly need to exercise (in a space emergency, for example), your potassium-starved heart will have a tendency to become slightly arrhythmic.

So living in microgravity weakens your heart, overtaxes your kidneys, causes bone loss, and makes you pee a lot. You know what that sounds like? It sounds like being old. Space flight makes you feel old. So much for those doctors who said that life in microgravity could be a "fountain of youth."

There are ways to circumvent, or at least minimize, the bodily effects of microgravity. Sticking to a fanatical exercise schedule (exercising up to three hours a day!) can help to mitigate *some* of the health problems listed above, but not all. Even if you were to exercise for six hours a day, you are by definition not exercising for eighteen hours a day, and in weightlessness not exercising really means *not* exercising. The good stress that exercise puts on your heart can never match the weakening effect of near-zero G, and as a result even the most strenuous orbital exercisers (often astronauts who have served in the Marine Corps) still have health problems when they return to Earth.

This implies that in a spacefaring society like the Twelve Colonies, a complete zero-G environment could quickly result in the development of "microcultures"—different classes of personnel based upon gravity tolerance. In the military there would the spaceship grunts who would remain on board forever, never setting foot on a planetary surface again (and who could therefore afford to become zero-G acclimated). There would also be the elite fighter pilots who would spend their days training in their spacecraft, emphasizing cardio workouts. One interesting by-product of this, though, is that better cardiovascular fitness yields a lower blood pressure. That's good, right? However, lower blood pressure makes it more difficult for a body to pump blood to the head during the high-G maneuvering of combat. The pilots would need G-suits similar to those worn by modern fighter pilots—clothing worn around the lower extremities that compresses the legs and rear end during high-G maneuvering, hence squeezing blood into the head. Another separate category would be the Colonial Marines, who would have to do both cardiovascular workouts and weightlifting on the chance that they would be needed to make planetfall somewhere. No organization, even one that places great emphasis on separation of rank, would find it easy to maintain efficiency with a crew divided into separate cultures as such. So it's almost imperative to create artificial gravity if you're going to create a civilization in space.

CHAPTER 13

The Wonderful World of Radiation

In science fiction TV and movies past, radiation is the great green glowing boogeyman that can do almost anything. In 1950s movies it fostered the growth of giant mutant ants, octopi, and grasshoppers. It created Godzilla (or Gojira), woke Gamera, attracted Kronos, and even gave a brachiosaurus the ability to distribute electric shocks. It gave Peter Parker spiderlike abilities, caused the hero to shrink in one movie and to melt in another. It influenced the genesis of the X-Men. There is perhaps no phenomenon in science that has been more misunderstood, or whose effects have been inaccurately portrayed, than radiation.

The general public has a bizarre love-hate relationship with radiation: we find it both incredibly cool and incredibly dangerous. We fry ourselves in tanning salons, then wear wide-brimmed hats and SPF 1,000,000 sunscreen outside. We protest against "irradiated food," then gleefully "nuke" something in a microwave oven. We've been conditioned to be wary of radiation by

Cylon Model Number Three D'Anna Biers.

Romo Lampkin.

both science fiction and real-world events like Chernobyl, Three Mile Island, and Hiroshima.

> **I don't want to be human. I want to see gamma rays, I want to hear X-rays, and I want to smell dark matter. Do you see the absurdity of what I am? I can't even express these things properly, because I have to—I have to conceptualize complex ideas in this stupid, limiting spoken language, but I know I want to reach out with something other than these prehensile paws, and feel the solar wind of a supernova flowing over me. I'm a machine, and I can know much more.**
>
> —John Cavil, Cylon Model Number One, "No Exit"

Yet radiation is essential to our very existence, and without it none of us would be alive. A cursory understanding of radiation would help us understand the physics behind some of the technology, weaponry, and astrophysical phenomena depicted on *Battlestar Galactica*.

Perhaps the main reason that radiation is so misunderstood is because several very different phenomena are collectively lumped into one term. Radiation can be separated into two different types: electromagnetic (EM) and particulate. The only real similarity between the two is that in each case something is being "radiated," or sent outward, but that "something" is very different in each case. EM radiation is energy in the form of waves emitted by an atom or molecule when it undergoes a transition from a high-energy state to a lower-energy state. Particulate radiation is the release of (usually) high-speed subatomic particles from unstable atomic nuclei, or from nuclear processes like fission or fusion.

To understand either type of radiation, it helps to first understand the basic model of an atom. An atom is the smallest unit into which an element can be subdivided while still retaining the chemical properties of that element. The ancient Greek philosopher/scientist Democritus observed that when a stone is split, the pieces have the same properties as the original rock. If cleft in two again, the smaller pieces are still rocks. Democritus reasoned that there was a limit to this progression, and eventually you would be left with pieces so small that they could not be further subdivided. He called these pieces "*atomos*," Greek for "indivisible."

Democritus further believed that atomos could not be destroyed and were unique to the material that they comprised. In other words, he believed that the atomos of stone were unique to stone, and were different from the atomos of other materials such as wood and water—a line of reasoning that was, in many respects, more advanced than that of later philosophers like Aristotle. It wasn't until the early twentieth century that science began to understand Democritus's atomos.

While an accurate image of an atom is incredibly complicated, for nearly 100 years scientists and teachers have relied on the Bohr Model, the simple model of the atom proposed in 1913 by the Danish physicist Neils Bohr, which itself was based upon the even-more-simplified Rutherford model, proposed by the New Zealand physicist Ernest Rutherford in 1911.

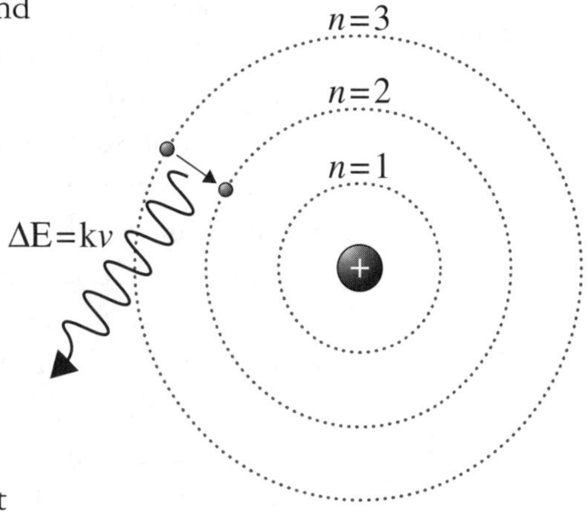

The Rutherford model of the atom.

Rutherford proposed that an atom was like a planetary system—with negatively charged "planets" (i.e., electrons) in orbit about a central "sun" (i.e., the nucleus). Just as the bulk of the mass of a planetary system is within the central star, the bulk of the mass of an atom is in the nucleus. The atomic nucleus consists of two different types of particles: protons and neutrons (collectively called nucleons). Protons and neutrons are similar in mass—a neutron has slightly more—but a proton has a positive charge and a neutron has zero charge.

An atom's chemical properties are a result of a value known as its atomic number—a count of the number of protons within its nucleus. Different elements have different atomic numbers. Hydrogen, the simplest atom, which comprises over 90 percent of the visible mass in the universe, has just one proton. Helium has two protons, lithium three, and so on. The number of neutrons in an atom's nucleus has no bearing on the chemical properties of that element, but as we'll see next, the count of neutrons can affect the way that atom behaves in other ways.

Planets are bound to their star by the force of gravity, but in an atom the force of electromagnetic attraction between negatively charged electrons and positively charged protons (usually) keeps electrons in bound orbits. For nearly all of the matter that you come into contact with on a daily basis—solids, liquids, and gases—the atoms have an equal number of protons and electrons. The positive charges of the protons equal the negative charges of the electrons; the atom is balanced, content, and dull. On those rare occasions when an atom has a net imbalance between its number of protons and electrons—when it gains or loses an electron—its behavior changes drastically. It becomes disruptive, a danger to all the other atoms around it. Ready at all times for the subatomic equivalent of a fight or a frak, such an atom searches desperately for the balance all the other atoms have, a balance it will achieve only at the expense of another. It becomes an ion.

The Bohr Model of the atom has yet another similarity with a planetary system: the distance between objects is vast. Just as the distance from, say, Earth to the Sun is much larger than the size of either object, the distance from the nucleus to the closest electron orbit is equally staggering. If the nucleus of a simple atom were the size of a grape, the nearest electron would be on the order of half a kilometer away. Since electron orbits of adjoining atoms don't overlap, electrons usually do not like to share their orbits with other electrons—a phenomena known as the Pauli Exclusion Principle, formulated by the Austrian physicist Wolfgang Pauli in 1925. Because of this, and because the distances between nuclei and electron orbits are vast, most matter with which we come into contact on a daily basis is just empty space. Remember the Second Law of *The Science of Battlestar Galactica*: "Space is mostly empty. That's why it's called 'space.'" It turns out that stuff is mostly empty, too!

In all of our discoveries of exoplanets—planets orbiting other stars—we've learned that planets can orbit at just about any distance from the central star. That isn't the case in an atom. To address difficulties with Rutherford's model, Bohr's model postulated that electrons can be found in only particular discrete orbits—which correspond to particular discrete orbital energies—around the nucleus. To move outward from the nucleus, an electron has to gain energy. To move closer to the nucleus, an electron has to lose energy.

There are many processes that can knock electrons to more distant orbits: collisions between atoms, electrical current flowing through a wire, even the absorption of light energy. Electrons, like Gaius Baltar, tend to like to stay at the lowest energy level possible; when they're externally energized, they will spontaneously, and usually immediately, drop back down to their original orbit.

In chapter 9 we saw that energy can neither be created nor destroyed. Since the outer electron orbit had a higher energy than the inner orbit, when an electron "drops down," where does that energy go? It gets radiated away from the atom in the form of a tiny packet of energy called a photon.

A photon is the fundamental "particle" of electromagnetic (EM) energy. EM energy comes in many "flavors," each comprised of photons with different energies. The entire range of EM radiation is known as the electromagnetic spectrum. EM radiation at the low end of the spectrum—whose photons have the lowest energies—is known as radio waves. The cell phone in your pocket is constantly radiating a small number of photons to various towers spread throughout your town. The word *RADAR* was originally an acronym that stands for RAdio Detection And Ranging. To determine the distance to other objects, a radar system radiates a pulse of radio photons, then measures the time it takes for that pulse to be reflected back. We can probably assume that Colonial and Cylon DRADIS systems radiate in this portion of the EM spectrum as well. (We'll elaborate on this much more in chapter 26.)

Photons that are slightly higher in energy than radio are called microwaves—the kind of radiation generated by microwave ovens when you "nuke" a burrito. The radar used by law enforcement to catch you speeding actually broadcasts in the microwave portion of the EM spectrum. Microwave transmissions are also used for satellite, spacecraft, and even Wi-Fi communications.

Still higher in energy is infrared (IR) radiation, sometimes called thermal radiation, which we normally associate with heat. Thermal radiation is emitted from objects when the energy of collisions between atoms and molecules is converted to electromagnetic radiation, with warmer objects radiating more energy than cooler objects. Any object whose temperature is above absolute zero (−459 degrees Fahrenheit) radiates photons in the infrared part of the spectrum. Yourself included.

There are many varied ways in which infrared radiation affects our day-to-day lives. IR radiation from an electric space heater can warm a room. Heat sinks on electronic components are designed to cool the devices by radiating heat away as thermal radiation. A tympanic ear thermometer determines a person's temperature by measuring the thermal radiation generated within the ear cavity. There have even been numerous instances over the run of *Battlestar Galactica* when we have seen Cylons fire heat-seeking missiles at Colonial Vipers and Raptors. Heat-seeking, or IR-seeking, missiles are programmed to home in on the thermal radiation emitted from the hot tailpipes of an adversary's spacecraft.

The spectrum of visible light ranges from red at the low-energy end of the EM spectrum to violet at the high-energy end. Many physical processes give off photons in this part of the EM spectrum—basically every method humans have used to create light, from fire to fluorescence. It's no accident that our eyes have their peak sensitivity at almost precisely the same wavelength as the Sun's peak visible light output.

Still more energetic than violet in the EM spectrum is ultraviolet (UV) radiation. UV radiation can be damaging to living tissue, and it is the primary reason why pale-skinned redheads fry like bacon at the beach. Although a type of oxygen molecule called ozone (O_3) shields Earth's surface from most of the UV radiation emanating from our Sun, enough gets through to be a concern.

At the upper end of the spectrum are very-high-energy forms of EM radiation: X-rays and gamma rays. As anybody who has ever had a dental exam knows, X-rays are used to look deep inside skeletal and dental frameworks. X-rays are generated by the radioactive decay of unstable atomic nuclei. They are also generated by astrophysical phenomena such as neutron stars, pulsars, and black holes. At the very-high-energy end of the EM spectrum are gamma rays—the same gamma radiation that mutated comic book (and movie) character Bruce Banner into the Hulk! Gamma rays are generated by radioactive nuclei, thermonuclear explosions, the nuclear fusion processes that generate power within the cores of stars, and supernovae. Gamma rays can do extreme damage to the cells of living tissue; parents, don't buy that gamma-ray generator on eBay thinking it'll make a good toy for the kids.

Visible light, at least visible for humans, corresponds to a narrow portion of the entire EM spectrum. There are many cosmic and astrophysical phenomena that are more clearly seen in portions of the EM spectrum outside of the visible, which explains why NASA's four Great Observatories—space-based telescopes—are geared to see in the infrared (Spitzer Space Telescope), visible (Hubble Space Telescope), X-rays (Chandra X-ray Observatory), and gamma rays (Compton Gamma Ray Observatory). Colonel Tigh alluded to this in the episode "Water" when he said, "Optical and X-ray telescopes say we've got five systems within our practical jump radius. All five have planetary bodies." (How would Colonel Tigh use X-ray telescopes to find planets? Repeat the First Law of *The Science of Battlestar Galactica* to yourself: "It's just a show, I should really just relax.")

This also clearly explains Brother Cavil's overwhelming frustration with his basic design, as well as that of all the humanoid Cylon models: they can't "see gamma rays" nor can they "hear X-rays." Keeping the basic human form constrains them to observing the same narrow portion of the EM spectrum as their Colonial counterparts. As machines, nothing would prevent them from seeing the entire EM spectrum—giving them the ability to observe the universe in a way that no human lacking special instrumentation ever could.

Photons propagate through a vacuum at 299,792,458 meters per second (a value for which we normally use the variable c). The key phrase is "in a vacuum." Photons travel more slowly through other transparent media (air, glass, and water being examples) than they do through a vacuum. The speed of photons in a vacuum is, as far as we know, the upper speed limit in our universe. Photons have zero mass and travel at c; any object that has mass must travel more slowly. We'll examine how this rule may be broken, or at least severely bent, in chapter 22.

The vehicle for electromagnetic radiation is the massless photon. The vehicles for other types of radiation are the many varieties of subatomic particles, all of which have mass. Generally these forms of radiation are generated as a result of processes within an atomic nucleus, or processes involving the interaction of multiple nuclei. To understand these forms, we must delve into the nature of the atomic nucleus just a trifle more.

The three most common forms of radiation emitted are called alpha, beta, and gamma radiation.

Alpha particles consist of two protons bound to two neutrons, and are equivalent to a nucleus of helium. When a radioactive element such as uranium gives off alpha particles, it's a sign that one of the atoms of uranium-238 has decayed into an atom of thorium-234. Alpha particles outside the body are relatively harmless; alpha particles inside the body are deadly.

Beta particles are nothing but free electrons.* When that thorium-234 atom subsequently gives off a beta particle, for instance, it becomes an atom of protactimium-234. Beta particles are slightly more energetic than alpha particles, but are also relatively safe as long as they're outside the body; inside the body they can rearrange DNA molecules to turn a harmless cell cancerous.

Gamma rays, as we have previously discussed, are extremely dangerous high-energy photons.

Another type of subatomic particle that can be radiated by unstable nuclei is the neutrino. Neutrinos travel very close to the speed of light, have an almost nonexistent amount of mass, and do not have an electric charge. Neutrinos do not easily interact with normal matter, so they can pass through ordinary matter unperturbed; they are also extremely difficult to detect. Every second, over 50 trillion neutrinos generated within the core of the Sun pass through your body. In addition to being by-products of radioactivity, neutrinos are also emitted in staggering amounts by nuclear fusion reactions—the kinds of reactions that power stars—and in even more staggering amounts when a star explodes in a supernova.

Finally, we return to the concept of the ion. Depending upon how it interacts with atoms, radiation can be classified as either ionizing or non-ionizing. Radiation, either particulate or electromagnetic, that has enough energy to strip electrons from an atom—enough energy to

---

*Or free positrons. They don't discriminate.

create ions—is called ionizing radiation. The three most common types of radioactive decay products—alpha, beta, and gamma radiation—are all forms of ionizing radiation.

Some forms of radioactive decay even emit subatomic particles called positrons. Positrons are not ionizing per se. They are electrons that have a positive charge instead of the negative charge to which we're accustomed. This means that positrons are a type of antimatter. Most of us have read or watched enough science fiction to know (or think we know) that when matter and antimatter come into contact, there is an explosion. That actually occurs only when large amounts of matter and antimatter come into contact, on the order of about one gram or so. When a particle and its antiparticle collide, the entirety of their mass is converted into energy (by $E = mc^2$) in the form of gamma rays—which are ionizing.

Ionizing radiation can do severe damage to living tissue by creating free radicals—highly reactive atoms or molecules that readily form chemical bonds and form numerous different compounds—within your tissues. While organisms require a certain level of free radicals to perform basic biological functions, when the concentration of free radicals gets too high, beyond a body's ability to regulate, bad things happen, like wrinkles, old age, and cancer.

Ionizing radiation can damage or alter the DNA sequences within an organism's cells. Changes to the genetic code of an organism are called a mutation, and any substance that causes the mutation is called mutagenic. Since DNA is the genetic code that, among many functions, tells cells how to divide, some mutations can have very profound effects. Most mutations are harmful, but normally a healthy body can either repair damaged DNA or kill an unhealthy cell through a process called apoptosis. On occasion, though, radiation can disable the DNA repair mechanism and block apoptosis. Any resulting mutated cells will pass their damage on to subsequent cell divisions, which in turn frequently lead to various forms of cancer. We'll further discuss the harmful effects of radiation in much more (gory) detail in chapter 14.

We've established that ionizing radiation can cause mutations within multicellular organisms, and that some of those mutations can be passed along to subsequent generations of cells. It turns out that

THROW 'EM OUT THE AIRLOCK

There are a lot of boards on the Internet where fans can discuss their favorite television shows. They're places to praise a particularly good episode, or to pan a lame one. One of the *Battlestar Galactica* episodes fans took particular issue with was the third season's "A Day in the Life."

Chief Tyrol had to fix a slow leak in a battle-damaged airlock. Because he saw an opportunity to spend some "quality time" (or what passes for quality time to a fleet on the run) with his wife, he assigned Cally to assist him. While the couple was working in the airlock, a malfunction triggered the inner airlock door to close, trapping the Tyrols with a slowly dwindling air supply as well as the prospect that their son, Nicky, might be a parent or two shy by day's end.

Apollo's plan to save the Chief and Cally is to blow the explosive bolts on the outer

Cally and Galen Tyrol in "A Day in the Life."

»»

airlock door. Explosive decompression of the remaining air in the airlock should then expel the Tyrols out of *Galactica* through space and into the open hatch of a waiting Raptor. They'll repressurize the Raptor, and then deal with the physiological effects on the Tyrols.

I honestly beleave it iz better tew know nothing than two know what ain't so.
—Josh Billings
Affurisms (1865)

The ambient temperature of space is not absolute zero. It is *almost* absolute zero. In the absence of a nearby star or other heat-producing object, space is about 2.7 Kelvins, or about minus 455 degrees Fahrenheit. If Cally's temperature is about 98 degrees F, then there is a more than 500-Fahrenheit-degree temperature difference between her body and space. There should be a lot of heat transfer going on, right? The laws of thermodynamics state that heat should flow from the hotter material (Cally) to the colder material (space) until both materials are the same temperature. Since there is a lot more of space than there is of Cally, every last bit of heat should be drained out of her, and both she and the Chief (who is a bit bulkier, but still tiny compared to the universe) should freeze solid pretty quickly, right?

Not so fast. There are three primary methods of heat transfer: convection, conduction, and radiation. Convection occurs only when a fluid is heated from beneath, such as a pot of water on the stove, or deep within stars and planetary interiors, so it's not relevant to our discussion of the Tyrols.

Conduction, the temperature equalization that occurs when two objects of differing temperatures are in contact, is the method of heat transfer we're most familiar with in our everyday lives. When you touch a hot stove with your finger, energy is conducted from the stove to your finger. Alternately, when you hold a cold can of beer or soda, heat is conducted from your hand to your drink. When Cally is blasted out of the airlock and floating in free space, she's hardly touching anything; space is close to being a perfect vacuum. Her 3,000 square inches of skin would touch at best around 50,000 atoms in space—about the equivalent of a large protein molecule. If there's little or no medium surrounding your body, there's very little conduction to carry away your heat.

That leaves radiation. When Cally and the Chief are in space, they will lose heat as it is radiated away from their bodies. This is a much slower process than conduction, and is certainly not instantaneous. The Tyrols certainly would not freeze in the few seconds they were exposed to the space environment. If Tyrol and Cally were left in space, their bodies would freeze eventually, but long after they perished by asphyxiation (as eventually happened to Cally, poor thing).

In the movies *Total Recall* and *Outland*, we see a less-than-accurate depiction of

››› the physiological effects of human exposure to a low-pressure environment. At the climax of the first movie, Quaid (Arnold Schwarzenegger) and Melina (Rachel Ticotin) are cast onto the surface of Mars without spacesuits. Their eyes bulge out of their heads, ostensibly from the pressure difference between that in their tissues and that of the rarefied Martian atmosphere. Even more dramatically, in *Outland* the viewer is treated to several instances where people, subject to a near-vacuum on Jupiter's moon Io, literally explode from the pressure gradient.

This simply would not happen either.

Let's say you were blown out into space, and let's assume you did not hold your breath. What would the air in your system do? Would it burst your bodily tissues and/or chest cavity, as in *Outland*? Or take the path of least resistance and flow out of your mouth?

The air in your body would mostly take the path of least resistance and burst out of your mouth. Your eyes, your intestines, your bloodstream, and the rest of your body would slowly outgas whatever air remained behind in your tissues. If you were blown into space—like Cally and Tyrol—the majority of the air in your body would simply flow from your lungs, out of your nose and mouth. As it passed from your body, cooling from the gas's sudden expansion might cause the moisture in your breath to freeze, and frost might form around your nose or lips. A small amount of air would take the more difficult route, and might burst small blood vessels in your nose and/or eyes, as well as a small fraction of the alveoli in your lungs. Nitrogen would bubble in your bloodstream. Your eardrums might very well rupture, but you wouldn't.

If, somehow, you managed to survive the experience, you'd have a major league case of decompression sickness, more commonly known as "the bends." The nitrogen dissolved in your blood as a normal function of respiration creates bubbles in the blood when your external confining pressure is reduced rapidly. This can cause incredible joint pain. Severe cases of the bends can lead to paralysis, even death.

A patient suffering from decompression sickness can be treated by placing them in a recompression or hyperbaric chamber. By overpressurizing the atmosphere around the victim, then depressurizing slowly, effects of the bends can be mitigated. This is exactly what happened for Cally, and in the same scene where Cally was relegated to a hyperbaric chamber, we also see that blood vessels in her eye did, indeed, rupture. One can infer that Cally had a worse case of the bends than did the Chief. Then, again, the Chief is a Cylon; who knows how resilient they really are?

In short, exposure to the near-vacuum of space would leave you very uncomfortable and unhappy, but if you were rescued quickly enough, you could easily survive it. ‹

if the cellular mutation occurs within the reproductive system of the affected organism, mutations can be passed along to descendants. More often than not, this is manifested as deformities—extra or misplaced limbs, albinism, missing organs, and the like. Maybe the old movies about mutant ants, grasshoppers, and spiders weren't so totally far-fetched after all.

CHAPTER 14

The Effects of Nuclear Weapons, or How the Cylons Can Reoccupy Caprica after a Few Days but Not Dead Earth after Two Thousand Years

Throughout most of the first-season episodes, we see Cylons living in the abandoned cities of a postnuclear Caprica. Then, in the fourth season, we learn that Dead Earth (nuked more than twenty centuries earlier) cannot be inhabited. Is such a thing possible? Can a planet like Caprica be nuked and then become almost immediately inhabited by radiation-sensitive people, while another, similar nuked planet remains a wasteland?

To understand this, we need to look at some of the effects of nuclear weapons. Giulio Douhet was an Italian general in World War I who practically invented the theory of strategic bombing from the air. In his 1921 book *The Command of the Air*, Douhet said that the perfect device for aerial bombardment of cities would consist of a mixture of high explosives to destroy buildings and create kindling; incendiary bombs to ignite that kindling and create a conflagration that would sweep across the area—a firestorm; and some form of long-term

Caprica being nuked.

poison gas to prevent firefighting, rescue, and cleanup forces from moving into the area.

Nuclear weapons essentially are General Douhet's perfect bomb. In a typical nuclear weapon detonated in the lower atmosphere, approximately 50 percent of the total energy of the bomb is spent as blast—Douhet's high explosive. Approximately 35 percent of the energy of a nuclear weapon is spent as heat or thermal radiation—Douhet's incendiaries. Approximately 15 percent of the energy of a nuclear weapon is spent as ionizing radiation and residual radiation—Douhet's long-term poison.

Let's look at blast damage first. Where Douhet imagined hundreds or thousands of medium-size conventional bombs raining down on a city, a single nuclear weapon can knock down buildings across an entire metropolitan area. In a "typical" strategic nuclear detonation, the blast usually happens in the air over a target, and is optimized to create maximum overpressure on the city itself, anywhere between

Admiral William Adama.

William and Lee Adama.

5 and 15 excess pounds of pressure per square inch. To create this overpressure, a 10-megaton nuclear weapon can be detonated at as high as 10 kilometers (6 miles) over a city (smaller nuclear weapons will, of course, be detonated at lower altitudes). In the miniseries, we see the Cylon attack begin with a silent nuclear detonation in the sky, visible from Baltar's house. Later attacks are also clearly airbursts: one gives a newscaster in the studio a brief moment to react to the flash before she is killed by the blast. That same blast traveling outward in a spherical shock wave at about 1,000 kilometers (600 miles) per hour, near the speed of sound, also wipes out a reporter in the field about 1.3 kilometers (1 mile) away from the detonation. Much of the devastation of a nuclear weapon is caused by blast, since most buildings are not built to handle the overpressure brought by the shock wave (witness how Baltar's house disintegrates around him).

A few seconds after the blast wave hits a given location, air comes rushing back in to fill the "vacuum," thereby creating a second, smaller pressure wave from the opposite direction. Both waves do a great deal of damage to structures and people.

About 35 percent of the energy of a nuclear bomb is released as visible, infrared, and ultraviolet light. Any object in a direct line of sight to the explosion will receive some of this light. Depending on the color and composition of an object, this light will be reflected, transmitted, or absorbed. For a one-megaton bomb, anything in a direct line of sight up to approximately 1 kilometer (0.6 miles) from ground zero will catch fire and burn. Many such fires, started in the blasted remains of buildings, will almost surely evolve into a firestorm.

A firestorm is a fire that is so intense, and burns for so long, that it creates its own wind system. Firestorms are not unique to nuclear weapons; they also occur in forest fires, and were a major consequence of the Allied conventional bombings of the German cities of Dresden and Hamburg in World War II. In a firestorm, a large conflagration on the ground creates a column of superheated air that can rise into the stratosphere. The rising air pulls in low-level air behind it, bringing in fresh oxygen to feed the fire; in a large enough firestorm, the winds can reach hurricane force. As fresh oxygen feeds the flames, the fire burns hotter, and the winds become more erratic. This can sometimes create small cyclones of fire outside the burn zone, allowing the fire to spread. A firestorm ends when weather conditions change, or when firefighters manage to control the periphery of the fire, or when there is nothing left to burn. A firestorm in a nuked city has the potential to destroy more buildings than the original blast. The "Lest We Forget" photograph, which *Galactica*'s pilots touch for luck on the way out of the ready room, almost certainly shows a Colonial soldier witnessing a firestorm on Aerilon.

A worldwide nuclear war, with resulting firestorms across the planet, brings another danger—the real possibility of "nuclear winter."

When the particles of soot from a firestorm make it all the way to the stratosphere, they can remain suspended in the atmosphere for many years and block a significant portion of sunlight from reaching the surface. This can have the effect of cooling the surface of the planet, so much that temperatures in post-attack summer might be colder than pre-attack winter. In such a scenario, crops will not grow, and billions of people not directly affected by nuclear blasts will suffer as the world's food runs out.

Yet despite firestorms and nuclear winter, probably the scariest effect of nuclear weapons is radioactivity, which accounts for about 15 percent of the total energy of the bomb. Radioactivity, as we saw in chapter 13, "The Wonderful World of Radiation," is an effect of the breakdown of an unstable atomic nucleus, causing the nucleus to give off particles and waves, which could take the form of helium nuclei (a.k.a. alpha particles), electrons, or high-energy photons.

Curiously, the initial burst of radioactivity associated with a nuclear detonation is probably the last thing you have to worry about during a large-scale nuclear attack. If you're close enough to be harmed by the initial radiation, you're probably close enough to be killed by the blast or thermal pulse anyway.

But then there's fallout.

Fallout is bad. Unbelievably bad.

Fallout is made of vaporized soil, vaporized buildings, vaporized people, and vaporized nuclear material from the bomb itself, created in the first seconds of a nuclear blast. These generally condense into microscopic radioactive particles in the explosion's mushroom cloud.

Some of these radioactive particles fall back to the ground close to the explosion. Some are carried by winds, bringing their radioactivity dozens or hundreds of miles from the site of the explosion. Others make it all the way up to the stratosphere, where they spread around the world before bringing their radioactivity back to the ground.

The human race could be expected to survive an all-out nuclear war if the sole effect of atomic weapons were blast. We *might* survive the resulting nuclear winter if the only effect of atomic weapons was blast and fire. It is entirely possible, though, to design nuclear weapons in such a way as to maximize the intensity and duration of their radioactive fallout—blanketing Earth with intense radiation that lasts for years, destroying not only humanity, but nearly every living creature on the land and sea.

This is probably the difference between nuked Caprica and Dead Earth: the fallout. The mechanical creatures who blew up Galen Tyrol at the farmers' market on pre-Dead Earth could easily have "salted" their nuclear weapons with a few kilograms of any of the following parent isotopes.

Parent Isotope[1]	Radioactive Product	Half-Life
Lithium Fluoride	Fluorine-18	109 minutes
Magnesium-24	Sodium-24	15 hours
Gold-197	Gold-198	2.697 days
Tantalum-181	Tantalum-182	115 days
Zinc-64	Zinc-65	244 days
Cobalt-59	Cobalt-60	5.26 years

The parent isotopes are non-radioactive, relatively abundant, and easy to obtain. A nuclear explosion will convert these parent isotopes to a specific radioactive isotope, and that's where the killing lies.

The radioactive half-life is the expected time it takes a radioactive substance to decay to half its initial strength. Suppose we have a quantity of radioactive cobalt-60 that is undergoing an average of 1,000,000,000 radioactive decays per second. A little more than five and a quarter years later, that same quantity of cobalt-60 will be experiencing only 500,000,000 decays per second. Five and a quarter years later (10.52 years after we started measuring), it will have an average of 250,000,000 decays per second, and so on. Since cobalt-60 radiates particularly energetic gamma rays, any area contaminated with the isotope is going to remain uninhabitable for a long time beyond the five-year half-life. By carefully exploding an awful lot of cobalt bombs in ways that will simultaneously maximize the amount of fallout *and* put that fallout into a planet's stratosphere, the mechanical creatures that destroyed the 13th tribe could, in theory, create a "doomsday shroud" of radioactive particles that would completely blanket Dead Earth and destroy all humanoid and animal life.*

Isotope salting works with "standard" nuclear weapons. But it is entirely possible to redesign nuclear weapons in ways that will modify their innate radiation effects. When we see the Cylons walking around undamaged portions of Caprica City soon after the attack, it leads to the assumption that—at least in certain areas of their attack on the Twelve Colonies—the Cylons used neutron bombs.

Neutron bombs, known in Pentagon-speak as "enhanced radiation weapons," were thought up by Sam Cohen, a researcher at the Lawrence Livermore Laboratories. During the Korean War, he had seen first-hand the destruction that war brings to civilian populations. He knew that there was talk in Washington of possibly using atomic weapons in Korea, and Cohen realized that if conventional war treated civilians so badly, nuclear war—which can be targeted almost exclusively at civilians—would be even worse. There had to be a way to use nuclear weapons in a way that would benefit, or at least not hurt, civilians.

Cohen used his nuclear physics background to develop the idea of a neutron bomb: a nuclear explosive specifically built so that most of the energy comes out as radiation—not blast or heat—and most of the radiation comes out as neutrons. A small neutron bomb exploded three thousand feet in the air will do minimal damage to buildings, but will put out enough radiation to kill nearly every person in a half-mile radius. Being neutral particles, neutrons have a strange effect on the human body: above a certain limit, you will certainly die relatively quickly; below that limit, you will certainly live, with very few side effects. Also, neutrons do not produce fallout or lingering

* Yes, this was more or less the idea behind the Doomsday Machine in the 1964 movie *Dr. Strangelove, or How I Learned to Stop Worrying and Love the Bomb.*

radiation to contaminate an area—an area blasted by a neutron bomb can be reinhabited a few minutes after the explosion (or after the bodies have been carted away). Cohen calls neutron bombs the most moral weapons ever made.

Could this be the difference between Caprica and Dead Earth? Did the conflict between the humanoid Cylons and their Centurion creations result in the use of cobalt bombs for at least some of the attack on Dead Earth? Based on D'Anna's decision to stay on Dead Earth to die, it would seem so; a place still radioactive 2,000 years later almost certainly had salted bombs used on it. Did the Significant Seven use neutron bombs for at least some of their attack on Caprica? Based on the Cylon presence in a relatively undamaged Caprica City soon after the attack, it would seem so. It makes sense. The Significant Seven Cylons have always been obsessed with the idea of moral behavior.

PART THREE

THE TWELVE COLONIES AND THE REST OF SPACE

CHAPTER 15

Our Galaxy

A galaxy is an ensemble of stars, multiple star systems, star clusters, nebulae, gas, and dust, bound together by gravity into a loose structure. The spectrum of galaxies ranges from dwarf galaxies with a few tens of millions of stars up to giant galaxies having upward of a trillion. The universe, for its part, contains hundreds of billions of galaxies.*

Because our sun Sol is in the Milky Way Galaxy, we can see our Galaxy only from the inside, as a creamy band of light that splits the night sky in two. Without our knowledge of astrophysics, ancient people were free to make up their own stories about what it was. The Dogon people of Africa saw the Milky Way as a spine, a backbone

*In the original 1978 *Battlestar Galactica*, the terms "galaxy," "universe," and "star system" were used willy-nilly, often interchangeably, endlessly confusing—or at least annoying—viewers who knew anything about astronomy.

holding up the skin of the night sky. The Lenni Lenape people of North America saw it as the smoke from the campfires of the braves who had gone over to the other side of death. It is the Greek image, however, that has stayed with us: the root of "galaxy" stems from their *gala,* meaning "milk," a poetic description of the river of white in the night sky. The first known use of the term "Milky Way" in English literature was in a poem by Geoffrey Chaucer.*

Approximately twenty-four hundred years ago, the Greek philosopher Democritus proposed that the bright band in the night sky might consist of numerous distant stars. Proof of this came two thousand years later when Galileo Galilei observed it with his simple telescope. For three centuries, scientists held that the Milky Way was the universe in its entirety. Hints to the contrary, however, came toward the end of the eighteenth century.

Charles Messier was a French astronomer interested in hunting comets. The predicted return of Comet Halley in 1759 set off a mania for comet discovery among the telescope-wielding natural philosophers of the day. Messier, still in his twenties, caught the bug. Comet-hunting was an arduous task, since undiscovered comets appear as dim, fuzzy spots in the telescope eyepiece. The problem was that telescopes in Messier's day were not very good by today's standard. Practically *any* object—galaxies, nebulae, globular clusters—could also appear as dim, fuzzy spots. In his hunt for comets, Messier repeatedly kept coming across the same fuzzy objects in the same parts of the sky. He began to catalogue these objects, not because he thought they were intrinsically interesting, but so he could remove them from his comet searches. In 1771 he published his first list of The Top 45 Annoying Objects That Are Not Comets (actually, it was called *Catalogue des Nébuleuses et des Amas d'étoiles* ["Catalog of Nebulae and Star Clusters"], which he later

---

*"Se yonder, loo the Galaxie,
Which men clepeth the Milky Wey"
Geoffrey Chaucer, "The House of Fame," c. 1380. http://www8.georgetown.edu/departments/medieval/labyrinth/library/me/chaucer/HF.html, lines 936–937.

Sam Anders and Kara Thrace.

Sam Anders and Kara Thrace.

expanded to 110 objects). Today his list is simply called the "Messier Catalog," and objects within it are given "M" numbers. The Lagoon Nebula (see chapter 24), for example, was given the designation M8; the nebula in Orion was called M42; and the "Great Spiral Nebula"* in the constellation Andromeda was christened M31.[1]

Many of the Messier objects are star clusters: home to thousands or millions of stars. Galactic, or open, clusters are a loose gravitationally bound collection of a few thousand stars, and are usually found within the disk of a spiral galaxy. They usually are sibling stars, all born at roughly the same time from the same protostellar cloud. The best example of this is probably the Pleiades,† a cluster of hot blue stars about 100 million years old, visible in the north autumn sky riding on the back of the constellation Taurus. Their connection is tenuous at best—a passing star, or even encounters between the stars of the cluster, can disrupt the gravitational bond, ejecting the siblings onto their own separate paths through the galaxy. Take a good look at the Seven Sisters while you still can; they won't be together in a few hundred million years.

Globular clusters are spherical, and consist of many millions of stars that are all much more tightly gravitationally bound. Globular clusters normally orbit within the galactic halo—an extended, roughly spherical, region surrounding a galaxy like the Milky Way—though they can occasionally be found in the population of stars within the disk or galactic bulge.

Based upon his observations of the "spiral nebula" M31 in 1917, Heber Doust Curtis estimated that it was 490,000 light years away. Based upon this, Curtis became a proponent of the "island universe" hypothesis, which maintained that spiral nebulae are actually independent galaxies outside our own.

Astronomers were not universally in agreement; however, the matter was settled conclusively by Edwin Hubble in the early 1920s

---

*Remember, we're talking about 1771. Back then, they had no idea the Andromeda Galaxy was a galaxy. To them, the whole fuzzy patch of light was called the "Great Spiral Nebula" in Andromeda.

†Aka the Seven Sisters of Greek mythology. In Japan, it's called "Subaru."

using the new Hooker Telescope at the Mount Wilson Observatory overlooking Los Angeles, the largest in the world at the time. Hubble observed that certain regular variable stars in our Galaxy could also be seen in other galaxies. By comparing the brightness of those faraway stars with nearby stars, Hubble calculated that M31—and many other galaxies—were too distant to be part of the Milky Way. He also measured the speed of these newly distant galaxies, and found that most galaxies are moving away from us (M31 being an exception), and the more distant galaxies are moving away faster. This implies that our universe is expanding.

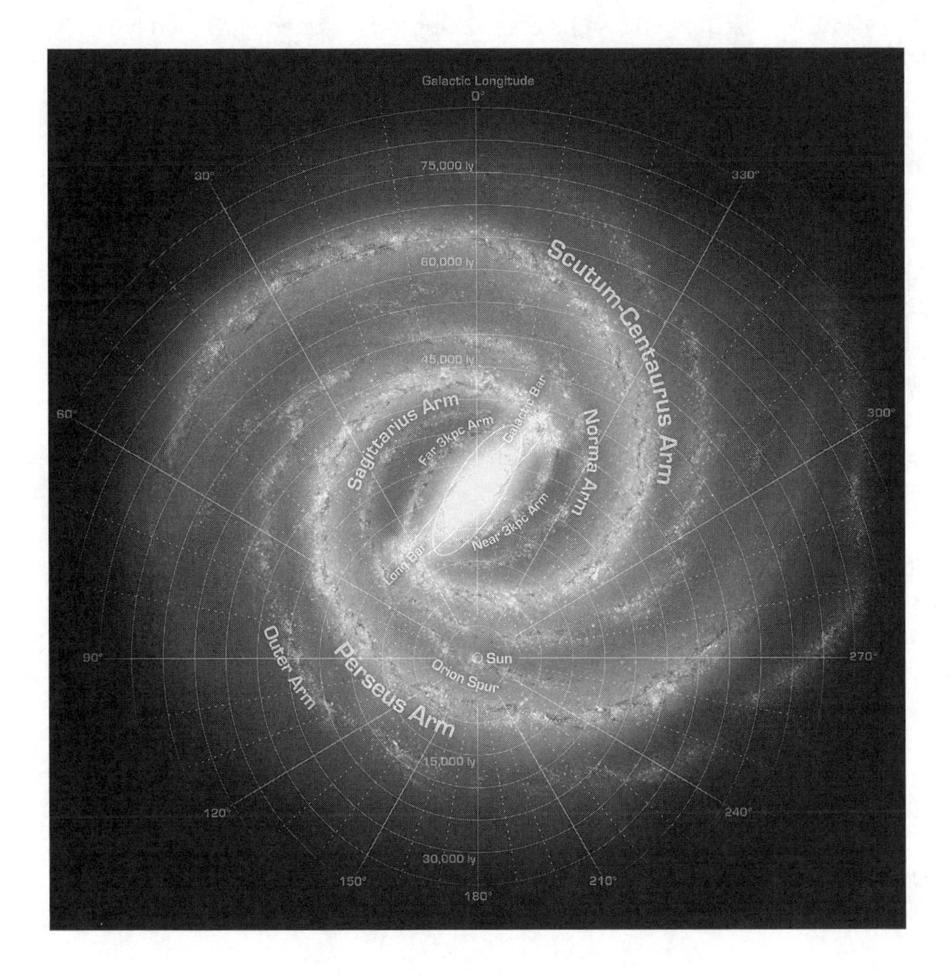

An artist's conception of the Milky Way.

PROTOPLANETARY DISKS: "THE HAND OF GOD" AND "SCAR"

The most famous one is perhaps the *Millennium Falcon* being chased by Imperial TIE fighters in *The Empire Strikes Back*—it's the science fiction staple called "dogfight in the asteroid belt," evading the enemy while massive boulders tumble in all directions, the slightest error meaning instant death, guaranteed to pump up the viewer's adrenaline.

In the first-season episode "The Hand of God," the original intent was to create a similar tone by using a similar setting. The Cylon tyllium mine/refinery was to be set on an asteroid immersed in an asteroid belt. The problem is, we only know one asteroid belt—our own—and the rocks there are much more widely spaced than is typically depicted in sci-fi shows. NASA routinely passes spacecraft through the solar system's asteroid belt—*Voyager*s I and II, *Galileo, Cassini, New Horizons,* to name a few—with little fear of collision. With rare exceptions, it's difficult even to see one asteroid from another in the asteroid belt.

There are other, more scientifically plausible ways to make this work. The mining colony could have been set in the ring system of a planet like Saturn (which has been done in both *Star Wars Episode II: Attack of the Clones* and *Star Trek: Voyager),** or even set in a protoplanetary disk.

The writers considered the latter option a better choice, but an interesting thing happened along the way—the emphasis of the episode changed. This happens fairly often with television episodes; what is considered "important" changes as the script is revised and improved. With all those changes, in the end we don't know where the Cylon tyllium mine/refinery was set and, in the end, it doesn't matter. Writers David Weddle and Bradley Thompson liked the protoplanetary disk idea, however, so it was recycled for the second-season episode "Scar."

Let's look at the formation of our own planetary system. (We'll go into more detail on this in chapter 16, "A Star Is Born.") As the cloud of hot gas circling a young Sol flattens into a disk, more material is "exposed," and radiates its heat off into space more readily and cools. Close to the protostar, where it is still quite hot, the first elements to condense out of the gas are metals, since they have the highest melting points.† As the disk cools, chemical reactions can occur, and dust forms.

Somewhat closer to the protostar is a place where it is just cool enough for ices of

*Episode 620, "Good Shepherd."

†A material's melting point and its freezing point are the same thing—we simply use one term or the other depending upon whether the substance is heating or cooling.

›››

»»

water and other compounds to form. This distance, different for every star, is called its *frost line*. Actually, there are several frost lines around each star, discrete places that mark the inner limit where ices of different compounds like water, ammonia, methane, and carbon dioxide can form.

As the disk of gas, dust, and ices swirls around the protostar, random collisions between particles occur. Occasionally they stick together, creating bigger particles. Occasionally these bigger particles collide and stick to form grains of material. Before long, the grains stick together to form pebbles, and before long the pebbles stick together to form boulders. Enough of these small collisions create solid objects known as *planetesimals*,* small planetary building blocks. Because of the protostar's heat, planetesimals formed near the star were made of nothing but metals and rocks. Out past the frost lines, the planetesimals were a mix of metallic, rocky, and icy bodies. There was actually more solid material from which to build planets in the outer solar system.

Planetesimals collided to form larger objects called protoplanets. Then something interesting occurred. As the larger protoplanets grew more massive, they developed a fairly strong gravitational pull. Collisions were now no longer random, but the larger protoplanets attracted others. This led to a scenario that the UCLA Professor William M. Kaula once termed "capitalistic growth" (that is, the rich get richer). This caused more collisions, making the protoplanets even larger, which increased their gravitational pull, which caused still more collisions, and so on. It is also at this point where we have an excellent setting for an episode of *Battlestar Galactica*: big chunks of space rock colliding with one another (and with any Vipers that might be in the area).❮

* Today planetary scientists use the term "planetesimal" as a generic name for both comets and asteroids—the leftover bits of planet formation.

Recent data from space-based observatories like the Hubble Space Telescope and the Spitzer Space Telescope tell us that our Galaxy is in the form of a spiral with two major arms radiating from a thick central bar, as well as several minor arms. The central bulge is approximately 16,000 light-years across. The entire Galaxy, from edge to edge, is 100,000 light-years across. In *Battlestar Galactica*, occasional mention is made of the "red line," the distance a Colonial ship can jump

A galaxy similar to ours: NGC 5866.

and still be reasonably sure it will end up at the desired location.* For a Colonial ship, the red line is on the order of five light-years. That means that, jumping as far as it reasonably can, it would take *Galactica* twenty-thousand jumps to cross the Galaxy—which would take a year and three months at thirty-three minutes per jump.

The pinwheel appearance of a spiral galaxy gives the impression that it is spinning, which, in fact, it is. At a distance of 26,000 light-years from the galactic center, the solar system orbits the center of the Galaxy every 225 million years. While that duration is a mere blink in the cosmic time scale, the last time the solar system was in the same place in the Galaxy, the dinosaurs were just getting a foothold on this planet, and Earth had only one continent.

---

*The difference between the final location and desired location for FTL jumps beyond the red line is more significant the farther the jump. One Raptor pilot alluded to this in the "Face of the Enemy" webisodes. When they realized that their Raptor had jumped beyond its red line, the pilot said, "The calculations just became nonlinear." Rephrased, that meant, "We're screwed."

Viewed edge on, a spiral galaxy appears much wider than it is thick, and it is.

Still, the disk portion of our Galaxy is roughly 1,000 light-years thick. It would take two hundred red-line jumps to cross it top to bottom. The central bulge is significantly thicker, about 5,000 light-years. Given these numbers, a rough volume estimate of the Milky Way Galaxy is *32 trillion cubic light-years*. Estimates of the number of stars in the Milky Way range from 200 billion to 400 billion.[2] If we assume that there are 400 billion stars in the Milky Way, then on average there is one star in every 80 cubic light-years. Although the stellar density varies greatly within the Galaxy, the average distance between stars is just over five and a quarter light-years. That distance is more than a single FTL red-line jump, meaning that it's not likely that a Colonial ship could jump from one star system directly to another. In less densely populated parts of the Galaxy, the actual distances between star systems might be much farther. This fact was echoed in the *Battlestar Galactica* series bible:

> *Galactica*'s universe is also mostly devoid of other intelligent life. Unlike [other science fiction series] crowded galaxies filled with a multitude of empires, ours is a disquieting empty place. Most planets are uninhabitable. . . When we do encounter a world remotely capable of supporting human life, it will be a BIG DEAL.*

Given the size of our Galaxy, it's not amazing that the Colonials took nearly three years to find Dead Earth. It's amazing that they found Dead Earth at all, even with the Prophecies of Pythia guiding them.

---

*This text, written by the series Executive Producer Ronald D. Moore, is the basis for the Second Law of *The Science of Battlestar Galactica*.

CHAPTER 16

A Star Is Born

Start about five billion years ago—about nine billion years after the Big Bang. We're in a nebula, a cloud of gas and dust, two thirds of the way out from the center of the Milky Way Galaxy. The cloud is vaguely spherical, about 480 trillion kilometers across.*

The cloud is big, but it's not very dense. The wisps of gas in this nebula are so sparse, you can almost see through the cloud without any problem. Still, 965 quadrillion cubic kilometers of gas—even thin, wispy gas—is going to have an enormous amount of mass. This particular cloud has the mass of approximately 100 Suns. But it has none of the heat of the Sun. The temperature of the gas in this cloud is about −450 degrees F, almost as cold as space itself, so cold that it has almost no heat at all.

This cloud is extremely fragile. It will break if something bothers it. If the cloud passes too close to a star, or if a star passes

*Light would take 50 years to cross from one edge of the nebula to the other. Another way to put it is to say that the cloud is 50 light-years in diameter.

too close to the cloud, or if a supernova explodes nearby, the cloud will either split apart into true nothingness, or it will collapse.

Let's watch it collapse.

Some small disturbance in nearby space causes a distortion in the cloud. It could be a change in temperature as the cloud absorbs energy from a supernova, or a fluctuation in gas density as the gravity of a passing star swirls the gas into vortices, or any number of things. The distortion affects only one part of the cloud, and that's enough. After a few thousand years, globules of gas, called dense cores, begin to form near where the cloud was disturbed. The cores are a little warmer than the rest of the cloud, but not by much, about -400 degrees F. It would be absurd to expect all the cores to be exactly the same size, and they're not. One core is larger than the rest, and so has more gravitational attraction than the others. The gravity of the larger core pulls the smaller cores toward it. Pretty soon, one dense core has eaten all the others, consuming as much as one-quarter of the entire nebula.

Perhaps the cloud was the gaseous remnant from the explosion of an earlier star, and perhaps it retained some of the spin, the angular momentum of its progenitor star. Perhaps the disturbance that caused the nebula to collapse imparted some spin. Either way, something has caused the cloud to spin. It starts slowly at first, but like a figure skater who spins faster the closer she pulls her arms into her body, as the cloud collapses, the faster it spins. The cloud also rapidly flattens, from a spherical shape to a disk with a central bulge—a very similar shape to that of our own Galaxy. The disk is called a *circumstellar disk* or *protoplanetary disk* (see the sidebar, "Protoplanetary Disks: 'The Hand of God' and 'Scar,'" in chapter 15).

This giant dense core, with as much mass as 25 Suns, is about one and a half trillion kilometers in diameter.* Its density is still very low, but the core's gravity strongly affects the rest of the nebula. Trillions upon trillions upon trillions of tons of dust and gas continue to fall from all directions toward the center of the dense core. As material falls from the outer fringes of the cloud toward the center under the

*Light would take about 2 months to cross from one edge of the core to the other. The core is 2 light-months, or ⅙ of a light-year, in diameter.

Roslin's assistant Billy Keikeya.

President Laura Roslin.

gravitational attraction of the core, its rotation speeds up. As the gas and dust speed up, the cloud's temperature increases, gravitational potential energy being converted to kinetic energy. Not all of the material in the circumstellar disk will fall into the central core; some of the material is orbiting too rapidly ever to fall in.

After about 100,000 years, the core's increasing gravity has pulled all of its gas and dust into a vague sphere about 16 billion kilometers across. That's still twice as big as our entire solar system. Light would take about fifteen hours to cross from one edge to the other. And with all that friction caused by the ceaseless gravitational pull, the core has warmed up to several thousand degrees. In fact, the core of the core, where the frictional heating was the highest, is giving off so much heat that the outward thermal pressure is starting to push against the inward force of gravity. The dense core is no longer just a dense core. It is now a protostar.

At this point, things start to happen pretty quickly. In only a few thousand years, the protostar collapses to about 450 million kilometers in diameter—slightly less than the orbit of Mars. The temperature inside the protostar is over 100,000 degrees Kelvin; the temperature outside is about 3,000 Kelvin, about half as hot as the surface of the Sun. The protostar is glowing red, but this red light is being produced by gravitational friction, not nuclear fusion, so the protostar is still not yet a star. To become a star, it has to contract even further. The protostar continues to collapse under its own gravity until it is smaller than Earth's orbit, then smaller than the orbits of Venus and Mercury, growing even smaller, hotter, and denser. Finally, somewhere in this last stage of contraction, the temperature of the core rises to a few million degrees Kelvin. The protostar has become a star.

Even if the mass of this new star were the same as our Sun, the brand-new star wouldn't be very Sun-like. Not yet. Newly born stars usually go through what scientists call the T Tauri phase of development.* In this phase, the newborn stars act more like adolescents: they are extremely

---

*Named for the star T Tauri, the next-to-last brightest star in the constellation Taurus the Bull. As you might expect, T Tauri is a newborn star, barely one million years old.

variable, being brighter on some days and dimmer on others, and their faces are covered in starspots. They are also "loud," in that they have a kind of superstellar wind that blows away much of the remaining gas in their surrounding nebula. This gas can be "vacuumed" up by large planetary embryos—planetary cores—in the outer system, eventually to form gas planets. T Tauri stars continue to contract, and after about 100 million years the core gets hot enough for hydrogen atoms to fuse into helium. This reaction gives off tremendous amounts of energy, in essence turning the star fully on. At this point, the T Tauri star settles down and joins the main sequence of stars.

Most of the stars *Galactica* visits—in fact, most of the stars in the Galaxy itself—are *main sequence* stars. After the hot volatility of the T Tauri phase, stars settle down to a more sedentary life, done with the giant flare-ups of adolescence. This lack of excitement during the main sequence inevitably makes science writers talk about "middle-aged stars." As that term suggests, main sequence stars are normal, middle-of-the-road stars. They exist because of the nuclear fusion of hydrogen into helium deep inside their cores. There's nothing too outrageous about their behavior.*

All stars have to put on a balancing act between the inward pull of their own gravitation, which aims to collapse all the material in the star into the smallest volume possible, and the outward push of the nuclear reactions going on in their core, which aims to blow the star apart. Main sequence stars have found that equilibrium point: the outward nuclear pressure perfectly balances their inward gravitational pull. Larger stars counterbalance their increased gravitational pressure with more active nuclear fusion cores; smaller stars under less gravitational pressure have less active cores. When the data are plotted on a graph, there is a nearly straight-line relationship between the star's mass and the star's luminosity. On the main sequence, larger stars are brighter stars.

---

*The opening narration to the 1955 science fiction masterpiece *Forbidden Planet* talks about "the main sequence star, Altair," perhaps as a way of falsely lulling us into believing that the place we're about to visit is normal.

This link between mass and luminosity also has consequences for a star's longevity as well. A star that's ten times the mass of the Sun isn't ten times as luminous—it is more than three thousand times as luminous,[1] and burns its fuel three thousand times faster than our Sun. Instead of living the 10 billion years our Sun is expected to live, the more massive star will burn itself out in only about 30 million years. It seems odd that a star with more nuclear fuel should have a shorter lifespan, but as with humans, packing on the pounds can take years off a star's life. Some of the more massive stars live for, perhaps, a million years. The smallest stars may live for trillions.

Astronomers classify stars according to both their total energy output and their temperature (or color). More massive stars are both brighter and hotter. Stellar surface temperature and color are, in some respects, the same measure. As an object heats up, it appears to change color. Anybody who has been around a blacksmith or a welder has seen this. The blacksmith places his cold metal into the fire and, as it heats up, it begins to glow red. Cool stars similarly appear red. As the metal gets hotter it glows orange, then yellow. The color corresponds to increasingly higher energies on the EM spectrum as it heats until it glows "white-hot" (white-hot metal actually has a bluish tinge, and very hot stars are similarly blue). In fact, the progression of colors is the similar to the order of colors on the rainbow: red, orange, yellow, and blue.* So the color red is associated with cooler stars; blue is the color of very hot stars.

From large-and-hot to small-and-cooler, stars are mainly classified by the letters O, B, A, F, G, K, and M.† Each spectral type is divided

* Where are the green stars? Stars that put out a lot of light in the green part of the visible spectrum *also* put out a lot of light in the blue and red part of the spectrum. To our eyes, roughly equal combinations of green, blue, and red light are seen as white light.

† Memorialized in the mnemonic "Oh, Be A Fine Girl/Guy, Kiss Me," which, if actually said anywhere outside of an astrophysics context will get you slapped with a sexual harassment suit so fast your head will spin. Alternately, if you wish to remember the sequence from cool to hot—M-K-G-F-A-B-O—then just remember the somewhat more disturbing mnemonic "Mickey Killed Goofy For A Body Organ."

into subclasses from 0 to 9, indicating where the star falls in its classification continuum: a G5 star, for instance, is halfway between a G0 and an F0 star.

Spectral Type	Color	Temperature(K)	Luminosity	Mass	Lifespan
O	Blue-Ultraviolet	> 28000	~1400000 Ls	~64 Ms	~ 500,000 y
B	Bluish White	10000–28000	~20000 Ls	~18 Ms	~32,000,000 y
A	White	7500–10000	~40 Ls	~3.1 Ms	~775,000,000 y
F	Yellow White	6000–7500	~6 Ls	~1.7 Ms	~ 2.8 billion y
G	Yellow	5000–6000	~1.2 Ls	~1.1 Ms	~10 billion y
K	Orange	3500–5000	~0.9 Ls	~0.8 Ms	~ 55 billion y
M	Red	<3500	~0.5 Ls	~0.4 Ms	~ 5 trillion y

It should make sense that smaller stars are far more common than larger stars, simply because they can be formed from smaller parent nebulae. M and K red dwarfs like nearby Proxima Centauri, Barnard's Star, or Wolf 359* comprise 93 percent of known main sequence stars. This implies that although Sol is often referred to as "average," it is actually larger than 95 percent of known main sequence stars. Large blue stars like Rigel and Sirius are comparatively rare and represent roughly only one in 3,000,000 known stars. The seven different types of stars have vastly different life spans, and vastly different fates.

A Type M star, less than one half the mass of the Sun, will end its life as a degenerate star† called a white dwarf. Over the hundreds of billions of years in its life span on the main sequence, the star will continue to undergo nuclear reactions in its core, converting its hydrogen fuel into helium until it runs out of hydrogen. Since the star is

* Made famous in *Star Trek: The Next Generation*, "The Best of Both Worlds, Part II."

† Not degenerate as in "pervert" (though there are certainly enough of those in the galaxy). Here degenerate refers to material that is made not of atoms, but of elemental particles: electrons, protons, neutrons, and quarks.

so small, it probably will not have the gravitational pressure needed to spark the fusion of helium. The star will leave the main sequence and will continue to contract until it is about the size of Earth. It will still glow from residual gravitational friction, but by itself it will never experience nuclear fusion again. Eventually, even its gravitational friction will cease. The white dwarf will cool even further and will eventually fade into darkness, becoming a black dwarf. At least, that's what we expect—the process of turning a white dwarf into a black dwarf takes so long that the universe probably isn't old enough for black dwarfs to exist yet.

A star larger than one half solar mass but smaller than about three solar masses—a star ranging from a smaller K to approximately a midrange type A—will also end its life as a white dwarf, but it will take an entirely different path to get there. When the hydrogen fuel in the star's core is all used up, the core will start to collapse gravitationally. This collapse has a strange side effect: it creates enough heat to initiate fusion in the layer just above the core, where there is still plenty of hydrogen. This layer of fusion provides energy to the rest of the star—enough energy to overcome the star's inward gravitational pressure. The outer layers of the star expand and cool until a new energy-gravity balance is reached. The star has now left the main sequence and become a red giant.*

After a few tens of millions of years, the hydrogen fusion layer burns itself out, causing the core of the star to collapse even further. In medium stars like our Sun, this can result in the *helium flash*, a momentary fusion of the star's helium core into carbon. (Yes, this is the same thing Lieutenant Gaeta yelled out just before the Algae Planet's star exploded. We'll explain that in a moment.) The flash lasts only a few seconds, but the energy pulse is eventually felt throughout the star. Over the course of hundreds of millions of years, the star will pulsate,

---

*When this happens to our Sun, about 5.5 billion years from now, the Sun will expand until it is approximately 170 times its current diameter. But don't worry. The Sun's loss of about one-fifth of its mass will make Earth spiral outward until it reaches a stable orbit approximately 1.4 times its current distance from the Sun. Earth will *probably* not be swallowed up by the red giant Sun. Mercury, however, is a goner.

shrinking and expanding as various layers of the star undergo nuclear fusion, then burn themselves out. Finally, the last burp of energy will blow off approximately half of the star's remaining mass, creating a thin shell of expanding gas—a planetary nebula. The remaining core of the star is yet another white dwarf.

A blue giant Type B or an even larger Type O star will end up as a supernova. After only a few dozen million years on the main sequence, a blue giant star will have used up the hydrogen fuel in its core. Much as the midrange stars did, the lack of core hydrogen will create a layer of hydrogen fusion just above the core. But these stars are so massive that the helium flash is actually not a flash, but a continued process of helium fusion. And once the helium fuel is used up, the temperatures inside the star are so great that the core begins to fuse carbon. Over time, the fusion process proceeds in layers throughout the star: an outer shell of hydrogen fusion, a shell of helium fusion below that, carbon below that, then oxygen fusion, neon fusion, magnesium fusion, all the way up the periodic table until chromium atoms in the core fuse with lighter elements to form iron.

This is where the star begins to die, because iron fusion does not release any energy. The core quickly runs through its chromium-to-iron conversion, and without any more fusion energy, it collapses in the blink of an eye. The rapid collapse of the star's core releases an enormous amount of gravitational potential energy in the form of heat, and the rest of the star explodes into superheated plasma moving at nearly the speed of light. The star has become a supernova.

There's an interesting side effect to this explosion: it is so hot that it can briefly cause iron atoms (and all the other atoms remaining in the star) to fuse into *every* naturally occurring element in the periodic table. Every element not already created in the star's shell—everything from the copper in our pennies to the gold in our bank vaults and the uranium in our nuclear reactors—is created during the supernova explosion and spewed out into space.

What about the core of the star? Does it also become a white dwarf?

No. The core of the star collapses on itself to the point where protons and electrons merge together to become neutrons. If the core

WHAT ABOUT ALL THOSE OTHER STARS?

Soon after the discovery of Dead Earth, Admiral Adama barked out an order to search for all nearby type K, G, and F stars. These are the types of stars most like our own Sun, a type G2. We would expect Kobol's sun to be similar, as well as those of the Twelve Colonies. These stars are relatively cool, and their initial nebula contained quite a bit of metals.* As a result, these stars have a relatively long life span, and are very likely to have solid planets. If you're looking for a place to settle humanoid life, you'll first look for a K, G, or F star.

Why would he eliminate the O stars? The A stars? All those M stars? It's because we would expect none of those classes of stars to have habitable planets. A good argument can be made that if life, especially intelligent life, is found in the cosmos, it will be around a star very similar in size, mass, and color to Sol. Sol-like stars are the kind for which Adama asked Gaeta to search.

* The term "metal," to an astronomer, means any element that is not hydrogen or helium.

Sam Anders on Dead Earth.

›››

›››

Naturally, the closer you are to a star, the hotter you are. The farther away, the cooler. There is a region around every star where a planet could have water in its liquid form—the planet is not too hot, it's not too cold, it's "just right." Not surprisingly, this region is called the star's *habitable zone* or *Goldilocks Zone*. For the solar system, the Goldilocks Zone extends from near the orbit of Venus to near the orbit of Mars. For small cool stars, like M and small K class stars, this zone is narrower and closer to the parent star. For larger hot stars—like O, B, and A stars—the Goldilocks Zone is not only farther away, but it is wider than that of Sol's.

Then why did Adama exclude these?

Although they have larger Goldilocks Zones, there are other factors that come into play when looking for a habitable planet. Earth II is 4.6 billion years old and according to the fossil record, it has had life for nearly that long. The oldest fossilized algae are 3.8 billion years old. Therefore it took 800 million years for life to appear on Earth. Stars in the large F to O ranges live a billion to a few tens of millions of years. That's simply not enough time for life to develop.

Most planets would need preexisting life to be habitable by Colonials. Kobol's (and Caprica's and Earth's) atmosphere was initially delivered to the surface by comets, so that atmosphere had a cometary composition: water, methane, carbon dioxide, and ammonia (among other compounds). That combination of gases would be deadly to humans, but the first life forms on Earth, cyanobacteria (or blue-green algae), literally ate it up. That algae inhaled the carbon dioxide–rich atmosphere and exhaled a gas that was, for it, a deadly metabolic toxin: oxygen. Oxygen is a highly reactive gas, and is extremely rare in nature in an unreacted form (that is, not bound to other atoms in a compound). As an oxygen-breathing species, humans would need to find a planet whose initial atmosphere had been "processed" by preexisting life. On a related note, if humans ever discover a planet with a high concentration of oxygen, life is a near-certainty.

So large stars simply do not live long enough to harbor human- or Cylon-habitable planets. What about the end of the stellar spectrum? What about the bulk of the stars in the Galaxy, the cooler M and K stars? These stars live extremely long lifetimes, so certainly they're better candidates for life. The key word is "better." They are still not "good" candidates. The Goldilocks Zones for cooler small stars are very close to the star, and also very narrow. This presents three problems. First, the mere fact that the habitable zone is small dramatically decreases the likelihood that a planet will form in the zone. Since the habitable zone of a small star is necessarily close, the gravitational pull of the star would tend to cause a planet at that distance to become *tidally locked*. This means that the same face of the planet would always face the star, like our moon—which is tidally locked to Earth—always presents the same face.

›››

»»

That means one side of the planet would roast while the other side froze. Finally, small stars tend to fire off more stellar flares. A stellar flare is a violent eruption in a star's atmosphere ejecting a stream of hot, highly energetic, charged subatomic particles into space. If a young planet were in the early stages of creating life-forming compounds, stellar flares could sterilize the planet's surface—especially since a nearby planet gets a bigger dose of radiation from a stellar flare than a planet farther away (like Earth), where the flare has time to dissipate.

Stars in the middle of the Main Sequence on the H-R Diagram,* the F, G, and K stars, survive on the main sequence long enough for life to form, they have moderate-sized not-too-close Goldilocks Zones, and they aren't as likely to spew stellar flares as smaller stars. So in the search for a habitable planet, Adama knew exactly where to look. ❮

* The H-R Diagram is a graph that displays the relationship between a star's luminosity and its temperature. With a few exceptions, most red stars are small and cool, most blue stars are large and hot, and most other stars fall in a continuum in between. The diagram was put together in the first decade of the twentieth century by the astronomers Ejnar Hertzsprung and Henry Norris Russell.

is between about 1.4 and 2.1 solar masses, the collapse stops there. The neutrons compact together until the core becomes essentially one giant neutron,* approximately 10 kilometers in diameter! A neutron star is born.

If the core is between about 2.1 and 5 solar masses, the collapse might continue until the neutrons break down into their constituent up and down quarks. The conversion releases an enormous amount of gamma radiation, and results in a star that is essentially one giant elemental particle, only a few kilometers in diameter.†

If the core is more than about 5 solar masses, the collapse will continue even past the quark stage, resulting in a *singularity*, or a *black hole.*

* The name for this material, neutronium, actually originated in science fiction, but has become the accepted term.

† Quark stars are still hypothetical at the time of this writing, but if they are true they might account for the missing "dark matter" in the universe.

CHAPTER 17

The Many Different Types of Planets

In *Battlestar Galactica*, the Fleet has passed through a number of star systems. The planets the Rag Tag Fleet has visited are a reasonable sample of the kinds of worlds we can expect to find as we explore the universe.

Planets

Name: The Twelve Colonies
Type: Terrestrial

The original *Battlestar Galactica* made reference to the fact that all of the Twelve Colonies orbit the same star: in the pilot episode, references are made to Cylon attacks of both the inner and outer planets. It strains credibility to expect there to be twelve habitable planets in the Goldilocks Zone of a single star, certainly not a G-type star like Sol or larger F-type stars like nearby Procyon. Hot O- and A-type stars have very wide habitable zones, but as we saw in the sidebar "What about Those Other Stars?" in chapter 16, these types of stars live very short lifetimes, sometimes barely long enough to allow planets to form. No, we're constrained to base the Twelve Colonies around F-, G-, and K-type stars. Even if we say two habitable planets are in mutual orbit, and that planets span the entire habitable zone, it's still difficult to think that the Twelve Colonies orbit a common star.

> **The beaches of Canceron are burning.**
> **The plains of Leonis are burning.**
> **The jungles of Scorpia are burning.**
> **The pastures of Tauron are burning.**
> —Cylon Hybrid, *Battlestar Galactica*, "The Plan"

How, then, can twelve human-habitable planets be packed into as close a space as possible to become the Twelve Colonies? The series bible for the reimagined *Battlestar Galactica* made a statement, with the same implication as in original series, with one major difference: the thirteen tribes traveled far away from Kobol, and eventually twelve of them settled in a star system with twelve planets capable of supporting life.

Although a lone star with twelve habitable planets in orbit may be a stretch scientifically, the term "star system" implies a small number of stars that are gravitationally bound. We can ask ourselves the broader question "What type of star system, as opposed to a single star, might be home to the Twelve Colonies?"

Half of the points of light in Earth's night sky are multiple star systems. While the vast majority of multiple star systems are binary stars—two stars in mutual orbit around their common center of mass—the nearest star to Sol, Alpha Centauri, is a trinary: Alpha Centauri A, Alpha Centauri B, and Proxima Centauri. Polaris, the North Star, is also a trinary system.

Alternately, perhaps the Twelve Colonies orbit a few different, albeit close, stars—stars that are a fraction of a light-year apart, such as the stars in an open cluster. Open clusters are physically related groups of stars held together by mutual gravitational attraction, typically with only a few hundred to a few thousand stars.

An open cluster is not, strictly speaking, a "star system," though. The Twelve Colonies are more likely to be orbiting stars in a multiple

Lee Adama and Anastasia Dualla.

Anastasia Dualla.

star system. Also, based upon what we've seen in the series and excerpts from the series bible, the colonies are probably closer than they would be were the planets orbiting stars in an open cluster.

The Twelve Colonies existed separately for most of their history, fiercely independent worlds with different cultures and societies. While they were clearly all linked together by heritage, they still found ways to war with each other, and presumably different alliances among the twelve rose and fell over the centuries according to the ebb and flow of history.

The Cylons were originally simple robots that grew increasingly complex with more and more powerful artificial intelligence. They eventually were used for dangerous work such as mining operations, and then they were used as soldiers in the armies of the Twelve Colonies.

Multiple star systems with more than two stars usually consist of a collection of binary pairs, or of binary pairs with one lonely star without a dance partner.

A trinary star system is statistically much less likely to form than a binary. A quadruple system is significantly less likely than a trinary. We know of a few star systems with six stars, but so far we know of none with twelve, which leaves us with the problem of how to pack multiple planets around a single star.

Computer orbital dynamic simulations tell us that 50 percent to 60 percent of binary star systems can form terrestrial planets within their habitable zones. But which habitable zones? Is the terrestrial planet orbiting only one of the two stars, largely unaffected by the gravitational perturbations of the companion star? Or is the planet in a distant orbit around both stars much in the way Proxima Centauri is around Alpha Centauri A/B?

Sol's Goldilocks Zone extends from near the orbit of Venus to near the orbit of Mars. With the right atmospheres, both could be habitable. So each star in a multiple star system could potentially have two or three habitable planets within the habitable zone of one of the stars' common center of mass. The Earth/Moon system has the largest satellite mass compared to the primary in the solar system, so high that a number of scientists think Earth and Luna form a binary planetary system. In the series *Caprica*, the colonies of Gemenon and Caprica are configured this way, revolving around each other as they orbit their star. The fairly sizeable Tauron presence in Caprica City also makes it plausible that that planet is in the same planetary system.

Yet it is still a very-low-probability event that a star system, even a multiple star system, would have twelve planets suitable for colonization. Nevertheless, we can apply known observable phenomena from physics and planetary science to postulate the scenarios in which this may occur. For the home of the Twelve Colonies we probably want a multiple star system consisting of Type F, G, or K stars having multiple planets orbiting each star. It's not impossible, but it is at this point that we must remind ourselves (1) that the hand of the divine has shown itself throughout the run of *Battlestar Galactica*, and (2) of the first rule of *The Science of Battlestar Galactica*.

Name: Kobol
Type: Terrestrial

Kobol, the homeworld of the Colonials, is a beautiful planet of oceans, clouds, mountains, lush vegetation, and a nifty holographic planetarium in a temple. It's also a planet in a stable orbit around a stable star, in a temperature zone where water can exist as a liquid, and it's loaded with the chemicals of life. This makes it a definite rarity in the galaxy; not one of the so-called terrestrial exoplanets discovered at the time of this writing is very Kobol-like.

How did Kobol come to be this way? Chances are, if it formed the same way Earth did, its atmosphere was brought by comets. But comets bring very little breathable oxygen; they tend to load up on things like carbon dioxide, water, methane, and ammonia. Why does Earth have a human-breathable atmosphere when its neighbors have atmospheres of carbon dioxide? Planetary scientists think that about 50 million years after Earth formed, it was struck by a planetoid roughly the size of the planet Mars. The impact blew off Earth's first CO_2-rich atmosphere, and the resulting splatter coalesced in space to form our Moon. Subsequent comet impacts and volcanic eruptions gave Earth its second atmosphere, which Earth's first bacteria could breathe. For a planet to be habitable, some mechanism has to keep the carbon dioxide content in its atmosphere low. This would be true for a planet like Kobol, as well as both Earths and New Caprica.

Name: New Caprica
Type: Terrestrial

New Caprica is a cold, barren, barely habitable world in the middle of a nebula. The planet might be located at the outer edge of its star's habitable zone, or the surrounding nebula might absorb too much sunlight to warm the planet—either way, only about 20 percent of the planet around the equator is suitable for colonization.

"Suitable" is a fuzzy word. A year after landing, New Caprica City is, at best, a shantytown. Just because you can set foot on a place and possibly raise a few crops doesn't mean that a planet has what is necessary to sustain a population, let alone the resources to let that civilization grow and develop. Like postattack Baltar himself, New Caprica seems to be a land of "just getting by," rather than a place of real growth.

One interesting thing is that New Caprica has an adequate amount of plant life, but no large animals are ever shown. Is the planet too cold for land animals to develop? Is it too young for land animals to have evolved? Or did the producers just not want to pay to both invent, then visualize, New Caprican cows and horses?

Name: Ragnar
Type: Jovian Gas Giant

Ragnar is a gas giant planet orbiting one of the stars of the Twelve Colonies. The Colonial government has placed a service and resupply anchorage (basically a space station in an impossibly low orbit) in the atmosphere of Ragnar. The location was chosen because the planet's radiation keeps the base hidden from Cylon DRADIS (Direction, RAnge, and DIStance). In the miniseries, Commander Adama discovers that Ragnar's radiation can, in time, damage the "silica pathways" leading to the brains of humanoid Cylons like Leoben.

Where other planets may have an atmosphere, a Jovian-type planet like Ragnar *is* an atmosphere. Most of it is, anyway. Except for a rocky metallic core, five to ten times Earth's mass, buried deep within the planet's center, a gas giant consists mostly of hydrogen and helium with traces of heavier molecules like methane, water vapor, and ammonia.

If a Jovian planet is mostly atmosphere, how would we measure its radius? Scientists have agreed on the "one atmosphere" rule: as you descend to deeper levels of a gas planet, the pressure of the overlying gases will naturally increase. At some point the pressure from the overlying gas is the same as the atmospheric pressure at sea level on Earth. At that point, measure the distance from the planet center, and you have just determined the "radius" of the gas planet. As with a solid planet, above the "radius" is the "atmosphere."

Name: Maelstrom
Type: Jovian Ice Giant

In the episode "Water," it was established that *Galactica* is able to resupply ships of the Fleet with water or fuel by piping these liquids directly from *Galactica*'s storage tanks into theirs. This procedure is called *underway replenishment* (UNREP). It's efficient to pump supplies directly from one vessel to another, instead of ferrying supplies incrementally using Raptors, but it leaves the Fleet temporarily vulnerable—it's nearly impossible for *Galactica* to maneuver when it's physically connected to another ship.

In the episode "Maelstrom" the Fleet is UNREPing again. As fate may have it, the Fleet finds a natural DRADIS jammer that allows them to hide from the Cylons while vulnerable: the Jovian planet we'll conveniently call Maelstrom. Maelstrom is intermediate in size and composition between Saturn and Uranus. "Ice giant" planets like Uranus and Neptune may have a solid core, but they are mostly composed of water, ammonia, and methane—compounds we've found to be common in comets. Hydrogen and helium are present in ice giants, but predominantly in the atmospheres.

All Jovian planets, gas giants and ice giants alike, have very powerful magnetic fields. When high-energy charged particles like fast-moving electrons become trapped and swirl within a magnetic field, they emit a form of electromagnetic radiation called synchrotron radiation. While it is possible to generate synchrotron radiation anywhere within the electromagnetic spectrum, from radio waves to gamma rays, the synchrotron radiation from Jovian planets tends to be in the radio and microwave portion of the EM spectrum. What types of technological devices send or receive information using radio or microwaves? Wi-fi networks do, as do satellite communications, electronic garage door openers, satellite TV and, yes, even radar (and presumably DRADIS). The ice giant planet Maelstrom was one huge natural source of electronic countermeasures.

Name: Dead Earth
Type: Terrestrial

It used to be such a nice place. Now parts of it look like winter in the Asiatic steppes, while other parts of it look like a bad day in Brooklyn.

Saul Tigh on Dead Earth.

Yet did you notice that even in a bombed out, radioactive wasteland, life still clings tenaciously to whatever it can? Wouldn't a postnuclear planet be absolutely lifeless? Not necessarily, although we can't be absolutely certain.

In 1986, an accident at the Chernobyl nuclear reactor, 70 miles north of Kiev, sprayed radioactive particles over parts of Ukraine and Belarus. The Ukrainian city of Prypiat, population 50,000, was evacuated 36 hours after the accident, when radiation levels in the town had risen to nearly one million times normal background levels. Though still abandoned, Prypiat and the surrounding forest contain a thriving ecosystem—a bizarre, mutant ecosystem filled with albino birds and giant radioactive ferns, but thriving nonetheless. Like Dead Earth, the region is still too dangerous for people to reoccupy, but it takes more than just a little radioactivity to wipe out all life (see chapter 14, "The Effects of Nuclear Weapons, or How the Cylons Can Reoccupy Caprica after a Few Days but Not Dead Earth after Two Thousand Years").

Name: Algae Planet
Type: Terrestrial

When the Colonial Fleet arrived at the algae planet they found an ocean full of photosynthesizing cyanobacteria (commonly called algae), a thriving ecosystem (developed just far enough to have developed algae and land plants), an atmosphere rich enough in oxygen to allow people to breathe unassisted (and, presumably, to provide an ozone layer), and a huge temple with an artistic rendering of an event that hadn't happened yet.

This is approximately the state of evolution that the Algae Planet was in when the Colonial Fleet arrived. Here on Earth, the development of green autotrophs led to an explosion of other life forms, specifically creatures that ate the autotrophs. This eventually led to the colonization of the land, because non-autotrophic life forms (a.k.a. "animals") now had access to these floating bags of high energy food. Perhaps the algae planet would have followed this same line of development, but since it was subsequently destroyed in a supernova explosion, we'll never know.

Satellites

Name: Ice Moon
Type: Ice Moon

According to Colonel Tigh, most planets are just "hunks of rock or balls of gas." But once you locate one of those planets—particularly the balls of gas kind—its moons are likely to be composed largely of ice, just like the one Boomer and Crashdown found at the end of "Water." Icy moons are not rare—at least not in our solar system—and it's reasonable to assume that icy moons, asteroids, and comets should be common in many types of planetary systems.

Of the nearly five hundred extrasolar planets, or exoplanets, discovered to date, most have been Jovian: Jupiter-like gas planets. Many of these orbit their parent star in extremely tight orbits, and are known as "hot Jupiters." In *Galactica*'s desperate search for accessible water, these types of planets would be the easiest types of planets to detect from a distance, but could also be instantly ruled out—they're located too close to their central stars for water to be in a solid state.

Though there are traces of ice on our own moon, the true ice moons in our solar system are found in the dim frigid realm of Sol's Jovian planets. We're therefore more likely to find similar icy moons in the middle to outer reaches of other planetary systems. There ices can condense around the solid body of a small moon. This further underscores an important point: what you call a "rock" depends upon where you are in the solar system. In the outer reaches of a planetary system, water ice is a naturally occurring crystalline substance, and is considered a rock by planetary scientists.

The three most well-known icy moons in our solar system are probably Europa, Enceladus, and Triton. Europa, the smallest of Jupiter's four Galilean satellites, was a shock to scientists when they first saw it close up during the Voyager flybys in 1979. Instead of a rocky, cratered moon, the probes sent back images of a smooth, cracked world of ice. Subsequent space probes have helped us to determine that Europa almost certainly has a small iron core, a rocky mantle, and a subsurface ocean of liquid water, capped by a crunchy frosted shell.

The problem with such ice, from a Colonial Fleet standpoint, is that it almost never is pure H_2O. Water on an icy moon in the far reaches of a planetary system, or orbiting a Jovian planet (or both), will almost certainly contain volatile contaminants such as ammonia, methane, or even other hydrocarbons like ethane. Fortunately, chances are that *Galactica*'s water purifiers will know how to remove methane and ammonia from the water. Small amounts of ammonia in the body are caused by normal protein breakdown. This ammonia is usually broken down into urea by the liver. People with liver problems, such as heavy drinkers, generally produce more ammonia in their urine than people with healthy livers. (Colonel Tigh, we're looking at you.) Then again, who knows if Cylon livers work the same way. Since the Cylons have made improvements on the basic human design in areas such as strength and stamina, wouldn't they have taken care of that pesky ammonia-in-urine thing? If so, there's an interesting test to confirm if Tigh is really a Cylon—get him drunk, then check his toilet.

The good thing about such ice, from a Colonial Fleet standpoint, is that it almost certainly contains volatile contaminants such as ammonia, methane, or even other hydrocarbons like ethane. As we have seen, water, methane, ammonia, and a source of energy (in the case of Europa, that would be frictional heat caused by Jupiter's gravitational pull) can, theoretically, combine to make the precursors of life. For Earthly scientists (and for Colonial fleets looking for algae-type foodstuffs), the most promising place to find extraterrestrial life in our solar system is not Mars, but rather Europa.

Name: Kara's Orange Moon
Type: Titan-like Moon

Kara Thrace, her Viper in a flat spin and damaged beyond control, ejects and parachutes onto the surface of a small, barren orange moon. The atmosphere is unbreathable, and a thick layer of haze hides the surface from space.

Sounds a lot like Titan.

Slightly larger than the planet Mercury, Titan is the largest moon of our own planet Saturn. It is the only natural satellite in the solar

Kara's orange moon.

system with a dense, stable atmosphere, and the only place outside of Earth that has large bodies of liquid on its surface.

How does a moon come to have an atmosphere? The planet Mercury has more mass than Titan, and therefore more gravity. How can a moon like Titan have an atmosphere, and the planet Mercury not have one?

Mercury is hotter. At Mercury's distance from the Sun, any gas atoms or molecules tempted to hang around the planet are heated by sunlight six and a half times more intense than at Earth. Those gas molecules heat up, speed up, and more readily achieve escape velocity. At Saturn, sunlight is 1/90th as intense as it is at Earth, nearly 1/600th as intense as at Mercury. Gas molecules don't move very quickly, and Titan's gravity has no difficulty hanging onto them in large amounts.

ORIGIN OF THE ALGAE PLANET

We'd just come to the point where we needed to get serious about how the rag-tag fleet would navigate toward Earth and he presented a whole PowerPoint presentation on space navigation, which gave us some ideas that the writers room twisted into "The Passage."

—Bradley Thompson, *Chicago Tribune* Online Interview

You never know what small thing a good writer will latch onto. In a room full of writers, especially those as talented as those on *Battlestar Galactica*, a single offhand comment can turn into an episode. Or two.

Before season three, the writer Bradley Thompson phoned and told me, "We're going to start getting serious about finding Earth this season. We need you to tell us what landmarks we have available to us." I threw together a half-hour presentation that I gave to most of the writing staff, speaking of celestial phenomena and observables like stars, pulsars, star clusters, nebulae, and black holes.

After the presentation I was waxing philosophical with the writers Bradley Thompson and David Weddle. I wondered aloud, "Stars much bigger than Sol don't live very long—bigger stars live by the motto 'Live fast, die young, leave a good-looking black hole.' It took life, the first algae, 800 million years to form after the formation of Earth. Stars much bigger than Sol live only, say, a billion years, not much longer. I wonder how many times in the history of our galaxy has life first appeared on a planet and barely had time to scream 'We're here!' before the Sun goes BOOM!"

Isn't that exactly what we saw in "The Eye of Jupiter" and "Rapture"?

While the existence of Titan's atmosphere isn't perplexing, its chemistry is. Mostly nitrogen, it contains a small amount of complex hydrocarbons ranging from methane (CH_4) to propane (C_3H_8), along with traces of other gases like carbon dioxide, hydrogen cyanide, argon, and helium. The thick orange clouds covering the moon probably come from the breakdown of methane gas by ultraviolet light from the Sun, and that's the problem. Ultraviolet light from the Sun, even the small amount available at the distance of Saturn, would completely break down all the methane in Titan's atmosphere

SERENDIPITY: A MANDALA-COLORED STORM

In the field of observation, chance favors only the prepared mind.
—Louis Pasteur

Serendipity. Look for something, find something else, and realize that what you've found is more suited to your needs than what you thought you were looking for.
—Lawrence Block

The hand of the divine may have been manifest throughout *Battlestar Galactica*, but occasionally it seemed to intervene in the production as well. The original plan was for the planet Maelstrom to have an extensive ring system like Saturn, in which the Fleet could hide while UNREPing. While the writers initially liked the idea, the attitude quickly turned to, "But didn't we already do that in 'Scar'?" Another way had to be found to hide the Fleet, and the convenient synchrotron radiation produced by a Jovian planet like Maelstrom proved to be just that. There was still a great reason to visit Maelstrom.

Wait, it gets better.

The executive producer Ron Moore wanted to set the action of this episode in the cloud layers of a Jovian planet because he envisioned Vipers and Cylon Raiders playing their game of cat and mouse, ducking in and out of different cloud layers—re-creating the feel of classic submarine movies like *Run Silent, Run Deep*. He couldn't have picked a better spot. Within Jupiter's atmosphere, like Earth's, there are clouds at different levels. Unlike Earth, where all the clouds are water vapor or ice crystals, having the same composition whether solid or liquid, Jupiter's cloud layers have different compositions and different colors. The outermost clouds are composed of white ammonia crystals. Impurities within the ammonia give these clouds a yellowish or beige tinge. Deeper is a layer of rust-colored clouds of ammonium hydrosulfide. Deeper still is a cloud deck composed primarily of water.

Wait, it gets even better.

The primary point of the episode "Maelstrom" is that this planet is where Kara Thrace finally meets her destiny—a destiny that, according to the Cylon Leoben, has "already been written." Again, a more perfect setting could not have been chosen.

If the planet is rotating, and Jovian planets tend to rotate very rapidly, the planet's clouds will separate themselves into bands called belts and zones; these counterrotating wind patterns run parallel to a planet's equator, and are not unlike the trade winds and westerlies on Earth. These belts and zones also segregate into clouds at different

›››

elevations. Zones have the higher clouds, and they are composed mostly of those white ammonia crystals; belts are deeper and are composed of rusty ammonium hydrosulfide clouds (we see this with Jupiter and with the gas giant planet in Kara's vision in "He That Believeth in Me").

Also, as on Earth, storms occasionally spiral through the tropics: the Great Red Spot on Jupiter—also known as the Eye of Jupiter—is a huge anticyclone, at least four hundred years old, and large enough to hold the four inner planets of our solar system! When the *Voyager 2* spacecraft passed Neptune in 1989, it saw a similar storm that was "only" the size of Earth that scientists called "The Great Dark Spot."

If Kara's storm on Maelstrom were anything like hurricanes on Earth, then peering into the eye of the storm, we could glimpse successively deeper layers of clouds. Framed within the nearly circular shape of the eye, the beige/yellowish white ammonia clouds would blend into the reddish ammonium hydrosulfide clouds. Looking deeper still, by the time we could see down to the clouds of water vapor, the light levels would be very low,

Starbuck's apartment on Caprica, the Mandala painted on her wall.

››› nearly extinguished. The clouds would tend to look bluish. In summary, we would see a storm with concentric rings of blue, then red, then yellow. Does this sound even remotely familiar?

Leoben was right. Kara's destiny had, in fact already been written. The painting on the wall in her apartment on Caprica—which we first saw in "Valley of Darkness"—turns out, entirely by coincidence, to be a painting of a cyclonic storm on a Jovian planet. Over her entire life Kara foresaw the place where she would meet her destiny. Serendipity. ‹

in a few tens of millions of years. Yet the methane is still there. How is that possible?

One possibility is that Titan has an active geology, complete with a warm interior that gives rise to volcanoes. Not the volcanoes we have here on Earth, but another type of volcanoes altogether. Titan has cryovolcanoes.

Here on Earth, a volcano is where incredibly hot liquid lava spews out of an opening in the planet's crust and eventually solidifies into rock. Titan is so cold that liquid water just a few degrees above freezing is equivalent to an exotic hot material like lava. Titan's cryovolcanoes spew a liquid mix (water, methane, and ammonia) out of openings in the moon's crust. These form clouds, which eventually rain (or drizzle) their hydrocarbons into the lakes of Titan. The rest eventually freezes into a solid material (ice).

On Earth a magma chamber is a subsurface cavity filled with molten rock under tremendous pressure. A volcanic eruption occurs when the molten rock is able to force its way to the surface. Consider a magma chamber on Titan. If ice is a rock in the outer solar system, then water can be considered magma. If cryovolcanoes on Titan spew water, ammonia, and hydrocarbons, and we recall that NASA's search for life in the universe begins with a search for liquid water, then a magma chamber on a moon like Titan would be a possible abode for life.

CHAPTER 18

Black Holes

In the episode "Daybreak, Part I," Raptor pilots Racetrack and Skulls, both participants in Felix Gaeta's mutiny, are paroled to undertake the dangerous mission of locating the Cylon Colony. On one sortie, they jump into what they initially think is an asteroid field. A sudden jolt and SINGULARITY DETECTED flashing on their DRADIS screen tells them that they're actually in an accretion disk of . . . something. Skulls calls it a singularity; Racetrack calls it a black hole. During a later briefing, the terms are used interchangeably: Starbuck calls the object a naked singularity, while Apollo says that the colony is "bound within the gravity well of a black hole." What is the difference? Is there a difference?

Black holes have been called many things: "a hole in space," "a monster that eats everything," "a sphere of no return."[1] They are almost always places where the collapsing core of a star that has gone supernova has attained nearly infinite density, creating a place where the

Admiral William Adama.

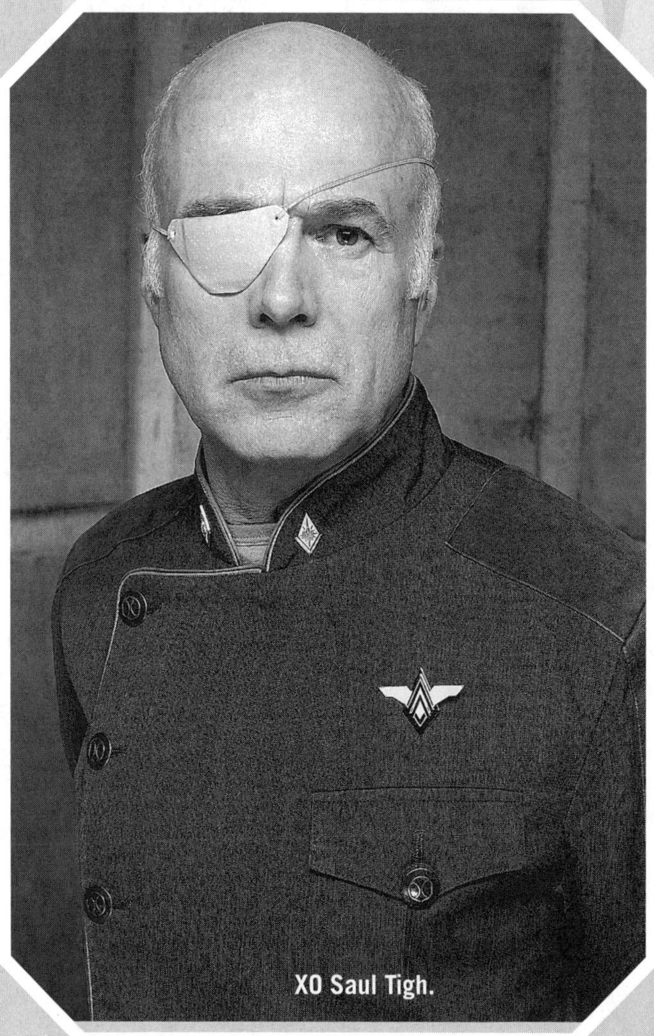
XO Saul Tigh.

mathematics of modern physics breaks down—a *singularity*. The Cylon Colony was in a stable orbit around a singularity. How is such a thing possible? Why wouldn't the Colony get sucked instantly into the black hole?

Contrary to popular belief, a black hole is not an insatiable vacuum cleaner, sucking in everything in the universe. If our Sun were magically replaced by a black hole of the same mass, Earth and other planets would continue in their orbits like before.* So the Cylon Colony and its Raiders, *Galactica* and her Raptors, could all fly around the accretion disk as long as they didn't get too close to the black hole. But how close is too close?

In the Eastern Front trenches of World War I, Karl Schwarzschild, a German astrophysicist serving as an artillery lieutenant, wrote a paper in which he calculated exactly what "too close to a black hole" means.† Black holes are called black because their gravitational fields are so powerful that near the singularity nothing, not even light, is able to escape. How close can an object get before it is forever trapped within the clutches of a black hole's gravity?

To be forever trapped by a black hole, an object would have to be unable to reach *escape velocity*. The escape velocity for a black hole—or any body like a planet or a star that has gravity—is simply a measure of the velocity‡ an object, like a spacecraft, must attain to break free of its gravitational hold, and is given by

$$v_e = \sqrt{\frac{2GM}{r}}\,.$$

* At least until an accretion disk formed, and started radiating energy.

† He sent his paper to Einstein, who was delighted with the findings. The two corresponded until Schwarzschild died of a skin disease he contracted on the Russian front.

‡ To a physicist, the concepts of speed and velocity are related, but not synonymous. An object's *speed* is a measure of how fast it is moving; its *velocity* is its speed in a given direction. With that definition, some claim that the term *escape velocity* should actually be *escape speed*, but to escape the gravitational pull of any celestial object the direction is implied—it must be radially outward.

Where v_e is the escape velocity, G is the universal gravitational constant, M is the mass of the body, and r is the radial distance from body center. Any NASA spacecraft bound for another planet must first escape Earth's gravity by reaching escape velocity, which is 11.2 km/s. For a black hole, this is obviously much higher.

The escape velocity for a black hole depends solely on the mass of the black hole, not upon the mass of the object trying to escape; therefore, a black hole's escape velocity is the same for a spacecraft, an atom, or even photons. The speed of light is a universal speed limit. If we determine the proximity to a black hole where the escape speed is equal to the speed of light, we have determined "the point of no return" or the point where nothing can escape.*

An examination of Schwarzschild event horizons for black holes of varying masses leads to a startling observation: Look at how small these event horizon radii are! Think about this for a moment. If our Sun were to become a black hole, you would have to be slightly less than 3 km away to get close to the event horizon.

Mass of Black Hole	Schwarzschild Radius/ Event Horizon
1 Earth	8.75 mm
1 Sun	2.95 km
4 Suns	11.8 km
10 Suns	29.5 km
100 Suns	295 km
1000 Suns	2950 km

But what if . . . ?

Recall that in "Daybreak, Part I" Starbuck said that *Galactica* would face a different jeopardy from the singularity the Cylon Colony orbited: "The tidal stresses are too strong, tear the ship apart before we got within 10 SU." A spacecraft near a strong source of gravity like a black hole that also has a large gravity gradient will experience a tidal force—when the gravitational force on the part of the spacecraft nearest the black hole is significantly greater than on the trailing edge. When you stand outside at noon, with the sun directly overhead, the sun's gravity is pulling on your head ever so slightly more than it is pulling on your feet. The difference is unnoticeable for something as small as a person,

*Unless, of course, it's FTL-capable. In the third episode of the series *Star Trek: Voyager*, *Voyager*, an FTL-capable vessel, is trapped within the event horizon of a singularity. No drama there.

but over the length of something the size of *Galactica*, the tidal effect of a black hole makes the ship even more fragile than you are!

What would happen as *Galactica* approached the event horizon of a 10-solar-mass black hole? Assume that the hull of the ship can withstand a force of 6,000 kilograms per square meter per second before it breaks apart. How close can *Galactica* approach before being torn apart? A Raptor? You?

How close to a black hole is too close?

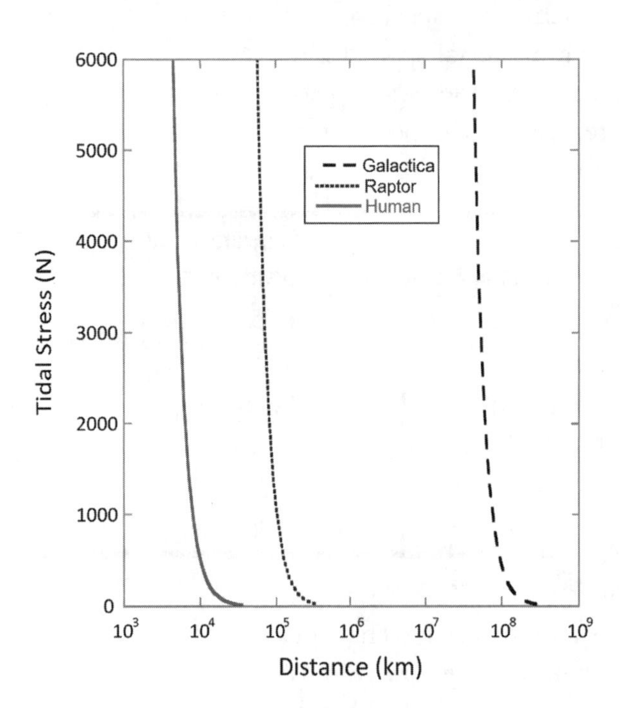

When approaching a black hole, remember this simple rule: a small object like a person can get much closer to a black hole before getting torn to shreds. In our 10-solar-mass black hole example, when comparing any two objects, the smaller object gets closer to the event horizon every time, yet all three objects undergo "spaghettifaction"—they are pulled apart lengthwise—long before reaching that point. Comparing the graph on this page to the Schwartzschild Radii in the table on page 179, we see that for a 10-solar-mass black hole, even something as small as a Raptor is torn apart by tidal stresses before reaching the event horizon. From this table we also see that Starbuck's comment about the ship being torn apart before getting within 10 SU was clearly hyperbole, even given *Galactica*'s compromised structural integrity.

As we saw in "Daybreak," that intense gravitational field of a singularity also creates an infalling spiral of matter and energy—an *accretion disk*. Recall that the Cylon Colony was in a stable orbit within the accretion disk surrounding a singularity. That orbit might best be described as "stable, with a lot of work." Gas drag, as well as the action of all those particles in the accretion disk impacting the Colony, would change its momentum and perturb its orbit in short order. Only by constantly compensating for that change in momentum could the Colony be said to have a stable orbit.

NAKED SINGULARITIES

We've discussed singularities, but Starbuck said that the object Racetrack and Skulls found was a *naked* singularity. What makes it different from other singularities?

Stars spin. Their cores spin. When a star explodes, the remnants of the star retain their spin, and so the resulting singularity also spins. It turns out that if a singularity is spinning fast enough, a relativistic effect called *frame-dragging* can leave it without an event horizon, exposed to the universe. This may not sound like a big deal until you realize that although the event horizon normally shields the universe from a singularity, it also shields the universe from the *effects* of a singularity. Within a singularity, matter becomes so dense, and the gravitational field so powerful, that the laws of the Standard Model of Physics, and the graceful equations that go with them, all break down. As the physicist Pankaj S. Joshi of the Tata Institute in India puts it, singularities* "are places of magic, where science fails."

Places of magic??? Where science fails?!?! Yes. At a singularity, the density of matter and the strength of gravity are immeasurable. Literally *anything could happen.* New universes could be created. Giant gravity waves could come crashing out. Trillions and trillions of bunny rabbits could appear. Protons and neutrons could simply cease to exist. Literally anything.

For the most part this is not a problem, since most singularities are surrounded by an event horizon. Anything on the other side of the event horizon effectively disappears from our universe forever, erased from our space-time as if it had never existed in the first place. The singularity's new universes, gravitational waves, and bunnies can never have any effect on our universe.

However, a naked singularity does not have an event horizon. You could fly right up to the edge of the singularity and fly back out again. Gravitational tidal forces permitting, of course. While you were there, you would be subject more to the laws of magic than of physics.

You might even come back with a new body in a brand-new Viper.

* We should mention that you don't *need* a collapsing supernova to create a black hole—what you need is some mechanism that can create the same incredible *densities* that are found in a collapsing supernova. It is possible that the Large Hadron Collider, a particle accelerator near Geneva, Switzerland, can create those densities when it slams subatomic particles into each other at enormous energies. One popular fear at the time of the LHC inauguration in the autumn of 2008 was that such experiments would create *mini–black holes* that would grow until they swallowed up the Collider, then Geneva, then Europe, and eventually Earth. If you're reading this book, that obviously didn't happen. Yet.

In the case of black holes, the accretion disk is formed by whatever gas and dust are available in the nearby stellar neighborhood,* pulled in to the black hole not only by its gravity, but by its magnetic field as well. These particles slam into each other as they spiral into the black hole, and the resulting friction makes them glow. As they spiral in toward the black hole, the particles of gas and dust get hotter and hotter until they begin to emit X-rays† just before they reach the event horizon.

Raptors are FTL-capable and, as such, could theoretically enter the event horizon of a black hole, and exit again. If somehow you could make it all the way through the event horizon and withstand the tidal stresses, what would you see?

Nothing. That is, nothing unusual.

You wouldn't necessarily know you had even crossed the event horizon. Light from the outside universe would still continue to reach you. It would be Doppler-shifted away from you, and it would seem to be less intense than it was, but it would be there until you reached the singularity, the place at the center of a black hole where the original matter of the star now has infinite density and zero volume. What happens when you reach that point? No one knows. However you managed to survive the tidal forces, you'd never survive the singularity. Is the singularity the end of your existence, or is it a portal to another universe? If it is a portal, and if you somehow could traverse it, would you remain who you are?

---

*The most interesting arrangement from an astrophysics point of view is when one member of a binary star pair turns into a black hole: if the other star survives the initial supernova explosion, and if it doesn't immediately get swallowed up by the black hole, then it will live a miserable life being slowly devoured, molecule after molecule, by its companion.

†Thus answering the obvious question "If black holes are black, how can you see them?" The answer is that we don't see the black hole itself, but a giant ring of material emitting powerful X-rays is going to be noticed. Some of the brightest X-ray sources in the galaxy are thought to be binary star systems in which a black hole swallows enormous amounts of gas from its companion star.

CHAPTER 19

There's No Sound in Space, and No Color, Either

It was one of the most thrilling moments of the show: Apollo is flying through the Ionian Nebula, ready to do battle with the Cylons, when suddenly Starbuck's Viper—which he had seen destroyed a few months before—appears next to his. Shaking off his shock and relief, Apollo manages to banter (and, predictably, argue) with Kara as they fly through the tendrils of colored gas in the nebula.

Scenes like this are ruining the field of amateur astronomy.

Amateur astronomers are people who study the night sky because they want to. They load up their cars and travel dozens or even hundreds of miles away from the light pollution of the cities and suburbs to a place out in the country with dark skies. There they unpack their telescopes—which they know how to dismantle and reassemble in the dark—and hold observing sessions, known as star parties, for the general public.

Amateur astronomers bring their telescopes wherever people have shown the slightest

Sharon and Karl Agathon.

Karl "Helo" Agathon.

interest at looking at the night sky: after-school events, overnight Scouting campsites, even daytime street fairs, where they swaddle their telescopes in insulation and filters to show off the Sun. At these star parties, children eager to learn about the wonders of the universe—and parents eager to recapture their own lost wonder—patiently wait in line for a chance to peer through the eyepiece. And after a few seconds of gazing at the glory of the heavens, they look up and utter the same three words: "Is that it?"

Their disappointment is the fault of one of the most successful satellites ever placed in orbit: the Hubble Space Telescope (HST). Launched in 1990 and put into working order in 1993, the Hubble can show us stars being born and stars dying; can show us planets around the Sun and planets around other stars; and can show us millions of galaxies shining like jewels in a patch of sky no bigger than your fingertip. The Hubble Space Telescope's beautiful images, swirling and burning with vibrant colors, have revolutionized the public's appreciation of astronomy. Mostly for the better, but sometimes very much for the worse.

To begin with, as this is being written HST's cameras are more than 20 years old, and thus are much more primitive than the one in your cell phone. That's worth repeating: the cameras in the Hubble Space Telescope are more primitive, and probably less powerful, than the simple camera in your cell phone. They don't even see in color. Instead, the image in the telescope is held very still while multiple black-and-white images are taken through different-colored filters. By taking a black-and-white picture through the red filter, the camera highlights the red objects in its field of view. Follow this with pictures taken through the green filter and the blue filter, and Hubble's cameras can capture all the necessary color information in a particular image. It is a simple matter to then combine the three filtered black-and-white images in a computer to make a full-color image. There are forty-eight filters in Hubble's eye, each tuned to a particular wavelength or band of light.

These filters are very important, because they can help astronomers figure out what is happening inside a nebula. A hydrogen alpha (Hα) filter, for example, shows scientists which areas of space contain ionized hydrogen gas. Ionized hydrogen gives off light in one wavelength: the rose-red hue of 656.281 nanometers, and the Hα filter is designed to block out other light and let only that wavelength through. That filter probably gets used a lot—since the material universe is mostly made of hydrogen, most nebulae, when seen in color, are overwhelmingly rose red.

But what about those Hubble Space Telescope images showing a full palette of Celestial Scarlets, Gaseous Greens, and Big Bang Blues? Those pictures that revolutionized the public's perception of astronomy. Those pictures that ruined any astronomical observation that couldn't give them a full-color "Hubble experience." Are the Hubble Space Telescope images *faked*?

No. Not exactly. Not faked. But they are enhanced.

Using color as a means of embedding some form of information onto an image is a relatively old practice, dating back at least to the earliest colored maps of Earth or fathom charts of the ocean. Since every element has a characteristic glow when ionized—ionized oxygen atoms glow emerald green, ionized helium glows yellow, ionized calcium glows dark violet—Hubble scientists use actual data and actual colors to present images that the human eye cannot see on its own.

Human eyes are practically blind to colors at very low light levels. The next time you see a bunch of amateur astronomers setting up camp, ask one of them to show you the Eagle Nebula (seen most easily in Northern Hemisphere spring). Because your eyes won't be able to see colors, even with the darkest skies and the largest amateur telescope, you'll be lucky to see something like a gray splotch of light, in the vague outline of an eagle in flight.

If you take a long-duration color photograph of the Eagle Nebula, either on film or electronically, the long exposure will help you to artificially collect more light and make the image brighter. Your resulting image might show something like a pink splotch of light, in the vague outline of an eagle in flight.

Then, of course, there's the iconic HST image of the Eagle Nebula, the one known as the "Pillars of Creation." The colors are not fake. Parts of the nebula with more ionized oxygen are colored green, as they "should" be. Other ionized gases are colored in their appropriate hues. By this method, HST scientists can subtract the "background noise" of the red hydrogen glow, and focus on the parts of the image that really show something interesting.

Before you start yelling that this is fraud, remember that any image you see through a telescope, like anything you see in a microscope or a spectroscope, anything from your child's ultrasound to your father's MRI, are all images that cannot be seen with the unaided naked eye. When your doctor adjusts the contrast on your ultrasound to show the outline of the fetus in more detail or more clearly, is the resulting image a fraud? Of course not. Hubble images use a similar technique to differentiate different types of material in a star cloud using intense versions of the real color that that particular material emits.

So what were Apollo and Starbuck flying through, and would it really look like that?

A nebula is essentially just a cloud of gas and cosmic dust in space. Sometimes, as in the (fictional) Ionian Nebula, the gas and dust were violently ejected by a dying star. In other nebulae, as we saw in chapter 16, the gas and dust are in the process of gathering together to create a new star, with some scraps of leftover dust gathering together under their own gravity to became the planets. Sometimes a nebula is both. As Carl Sagan and Moby made very clear, we are all made of the same material as the stars.

When the gas given off by a supernova collides with other gas within the star's planetary system (or even interstellar gas), the shock heats the gas to nearly 10,000,000 degrees, causing the nebula to glow. Red. Vivid red. Rose red. The 656.281-nanometer red of hydrogen alpha. Sure, there might be some other colors, but the predominant color would be red. So why were Starbuck and Apollo flying through pastel-colored veils of gas?

The simple answer is because they're on a TV show and that's what *we* were expecting. Had Starbuck and Apollo been flying through ghostly gray clouds of gas (as it most likely would have seemed to their naked eyes), or through uniformly red clouds of gas (as is more "accurate"), our attention would have been unnecessarily drawn away from their conversation, or the wonder of Starbuck's return. Instead, millions of viewers would have been wondering, "Why are the clouds only gray (or red)?" Thanks to the Hubble, everyone "knows" that interstellar clouds come in more colors than a box of 64 crayons, and, like a newbie at a telescope, we would have been jarred if the visuals had not matched our expectations.

This is where Moore's Law is at its finest. The point of *Battlestar Galactica* is to tell a story, not to present a scientific documentary. If it works better, in terms of interest and excitement, to have Starbuck emerge from the colorful clouds of a giant nebula in space, that's what they going to do. It's time to evoke the First Law of *The Science of Battlestar Galactica*: "It's just a show, I should really just relax."

CHAPTER 20

Water

Think back to the second episode of the first season. Commander Adama had spent the previous few minutes explaining to President Roslin that *Galactica*'s water recycling system is close to 100 percent efficient. He adds that since about one-third of the other ships in the Fleet were not built for long-term voyages, *Galactica* has to supply water-recycling services for them. As the president watches, *Galactica* begins the process of swapping dirty water for clean water with the *Virgon Express*. Then it happens. There's a dull explosion that causes the ship to rock. The lights start to flash in the CIC, and Lieutenant Gaeta yells, "Decompression alarm!" An outside view shows geysers of water erupting from holes in *Galactica*'s hull, boiling away to ice in the vacuum of space.

Later, at a briefing for Roslin and Adama, Gaeta reports that they've lost about 10 million "JPs" of water, about 60 percent of the ship's holdings. Baltar reports that the Fleet's population of 45,265 needs about 2.5 million "JPs" of water per week. As for replenishing

that supply, Colonel Tigh reports that there are five planetary systems within the Fleet's practical jump radius, but he doesn't hold out much hope: "Most planets are just hunks of rock or balls of gas. The Galaxy's a pretty barren and desolate place when you get right down to it." (Remember the Second Law of the *Science of Battlestar Galactica?* "Space is mostly empty. That's why it's called 'space.'") And it soon looks as if the Colonial civilization will not end with the bang of Cylon nukes going off, but with the whimpering croak of a parched throat.

There's a lot going on in the first few minutes of this episode. First, what is "JP"? If we assume that Baltar is telling the truth,* each Colonial needs about (2,500,000/45,265) = 55.23 JPs of water per week. If we assume that a Colonial week is the same as ours, that works out to 7.89 JPs of water a day; call it 8 JPs. NASA reports that astronauts on the International Space Station [ISS] are using about 12 liters of water per day. If we assume that an on-the-run civilian Fleet will use water as carefully as our astronauts, then 8 JPs is equivalent to 12 liters, and therefore each JP is about 1.5 liters. If that is the case, then *Galactica* holds nearly 16,666,667 JPs, or 25,000,000 liters of water, the equivalent of 10 Olympic-sized swimming pools. To refill 60 percent of that would require a ball of water 30 meters in diameter—the equivalent of about 7.5 million two-liter soda bottles.

Although Colonel Tigh was right about the composition of most planets, his statement was also misleading. There is more water in space than most people imagine, even if that water may be all but impossible to extract or utilize.

Water is actually one of the more abundant compounds in space—it's not everywhere, but not rare, either. In March 1969, when Dr. A. G. W. Cameron of Yeshiva University reported in the journal *Science* the presence of microwave emissions from excited water vapor in space, the nascent field of astrobiology probably gained its strongest foothold to respectability. Before that time, mainstream scientists thought outer space was dry—and why would anyone look for life in a waterless environment?

*Always an iffy proposition, though he has no particular reason to lie in this case.

Felix Gaeta.

Brendan "Hot Dog" Costanza.

This argument shows how chauvinistic we are toward water. Water is so closely associated with life here on Earth that for centuries we naturally assumed that life couldn't exist without it. Finding water in molecular clouds in deep space meant that life stood a chance of being nearly *anywhere* in the universe. Although it is theoretically possible for alien life forms to use fluids other than water, none of those fluids match the utility and ubiquity of H_2O itself.

Why? What's so special about water?

We've already defined life as a self-sustaining chemical system. Such a system needs a medium in which to sustain itself. It helps if that medium is a liquid, since liquids can easily transport substances like food and wastes. It helps if the medium remains liquid over a fairly large temperature range.[1] It also helps if that medium can dissolve a great many different chemicals, so that a living organism can make the greatest possible use of its available resources. For life on Earth, water meets all these requirements. Of course, it's a bit of a chicken-and-egg cycle: we happened to evolve in a star system in which water was abundant, so naturally we use water as our liquid medium.

It was exactly the galactic ubiquity of water that fueled one of the greatest complaints about this episode on BSG Internet fan boards: "Why didn't they just find a comet? They'd be done with all their water problems for the foreseeable future!" That may be true in theory, but in reality, our understanding of comets has come a long way since the American astronomer Fred Whipple called them "dirty snowballs."

Although there may be trillions of comets in the outer reaches of a planetary system, they are spaced widely and difficult to find and usually keep their water well hidden. The space probe *Deep Space 1*, in its 2001 encounter with Comet Borrelly, found nothing but a hot and dry surface without any obvious traces of ice. *Deep Impact*, a ballistic probe that smashed into Comet Tempel 1 in 2005, found the same thing on the outside—dry dust—with water existing only deep inside the comet. This dusty coating means that comets are dark objects when seen against a black backdrop of space. They're difficult to spot.

Further, if *Galactica* could somehow snare a comet, mining that water would require drilling. *Galactica* probably has access to drilling equipment,[2] but why go through all the trouble if you don't have to? When you're looking for easy water, comets might not be the way to go.

Well, then, where else can we look for water in a planetary system? When we look at the major bodies of our solar system, we find that Mercury,[*] Earth, Luna,[†] Mars,[‡] Phobos,[§] Deimos,[**] Ceres,[††] Jupiter,[‡‡] Europa,[§§] Ganymede,[***] Callisto,[†††] Saturn,[‡‡‡] Saturn's rings,[§§§] Enceladus,[****] Tethys,[††††] Dione,[‡‡‡‡] Rhea,[§§§§] Uranus,[*****] Neptune,[†††††] and Triton[‡‡‡‡‡] all have water, water vapor, or water ice in quantities that would allow everyone in the entire Colonial Fleet to take giant Japanese group baths in tyllium-powered hot tubs.

The water is usually on the surface in the form of ice, and therefore is relatively easy to access. If our solar system turns out to be average, then the Colonial Fleet should have no problem finding water in just about any other planetary system, on the surface of some icy moon.

Which is exactly where they find it.

---

* Mercury??? Hottest planet in the solar system? Closest planet to the Sun? *That* Mercury? That Mercury has *ice*?? Yes it does, in permanently shaded craters near its north pole.

† Found in the form of ice, under the dirt of the Moon's south pole.

‡ Ice is all over the planet, but mostly at the south polar region. There's also some evidence of liquid water underground.

§ This moon of Mars is loaded with veins of ice snaking underground.

** See Phobos.

†† This was the first asteroid discovered, in 1801. It was easy to see because it *is* covered in ice.

‡‡ Huge amounts of water are found in the planet's atmosphere.

§§ A moon of Jupiter with an ocean of liquid water capped by a crust of ice!

*** Another moon of Jupiter, with large swaths of ice on its surface.

††† See Ganymede.

‡‡‡ See Jupiter.

§§§ The rings of Saturn are largely made of ice or, like Ceres, rock coated in ice.

**** This moon of Saturn has ice geysers!

†††† Named after the Greek goddess of fresh water, this moon of Saturn is made almost entirely of ice!

‡‡‡‡ Another moon of Saturn, about 50/50 rock and ice.

§§§§ Another moon of Saturn, about 25/75 rock and ice.

***** See Saturn.

††††† See Uranus.

‡‡‡‡‡ This moon of Neptune has ice geysers!

PART FOUR

BATTLESTAR TECH

CHAPTER 21

The Rocket's Blue Glare: Sublight Propulsion

It most likely happened when some unknown-to-history Chinese philosopher improperly filled a bamboo tube with sulfur, charcoal, and dried pig urine crystals. Previous philosophers had found that this mixture of materials burned fiercely. When confined in a tube or jar, it exploded. Our unknown philosopher was probably doing just that—building a firecracker to make a huge bang in the hopes of scaring away evil spirits before a feast or other ceremony. It's not hard to imagine that that day, somewhere in the Celestial Kingdom, the bamboo tube caught fire on only one end. Instead of exploding, the burning material released rapidly expanding hot gases that sent the tube zooming in the other direction. The first rocket was born.

In a land beset by near-perpetual warfare, the idea of a self-propelled burning projectile must have seemed like the ultimate wonder weapon. The first rocket attack in recorded history took place in the year 1232 CE when Chinese defenders repelled Mongol raiders at the battle of Kai-Feng-Fu. The Mongols were fast learners.

Nine years later, when the Mongols were invading Europe, they used rockets of their own in the siege of Budapest. In 1258, the Arab world was introduced to rockets when the Mongols attacked Baghdad. The Arabs quickly added the rocket to their arsenal, and used them against King Louis IX's French army during the Seventh Crusade in 1268.

In 1650, the first European book on artillery was printed, and nearly 150 years later the Sultan of Mysore in India used iron rockets against the British East India Company troops. The British military engineer William Congreve reverse-engineered these rockets and made them part of the British arsenal by 1803. And the Congreve rocket had something new—a long stick coming out of the end to provide some measure of stability in flight.* Without the stabilization that the stick provided, a rocket was apt to land literally anywhere, even on one's own territory. By 1812, the British were so enamored of this weapon that the ferocity of their missile attack on Fort McHenry in Baltimore harbor led Francis Scott Key to write a poem expressing his pride at how the flag of the United States stood up to the rockets' red glare.

Until the twentieth century, rockets used solid propellant, usually a form of black powder or gunpowder. Solid propellants are the simplest form of rocketry, and they offer great reliability with very little velocity control. Solid rockets can't easily be throttled in real time the way a jet engine can: they either go, they don't go, or they explode.

But a solid rocket can't burn just any fuel. Ideally, the fuel must also carry its own form of oxygen. Gunpowder, a mixture of sulfur, charcoal, and potassium nitrate, gets its oxygen from the nitrate. The space shuttles' Solid Rocket Boosters (SRBs) use aluminum and butadiene as their fuel and ammonium perchlorate as their oxygen source, a combination called ammonium perchlorate composite propellant, or APCP. As the fuel burns, the oxygen is chemically released from its source material. This feeds oxygen to the burning fuel, which helps the fuel to burn better, which releases more oxygen, which burns more fuel, and so on. While substances like wood or coal burn very well, they don't have their own oxygen sources, and without that you're

---

* They were basically giant bottle rockets.

Kara "Starbuck" Thrace.

Starbuck and President Laura Roslin.

not going to find wood- or coal-powered spaceships outside of mad steampunk fiction.

Even though gunpowder works as a better rocket fuel than coal, it certainly isn't the best fuel. It was this quest for more energetic fuels that led to the second great development of rocket power: the liquid-fueled rocket in the early twentieth century. The development of liquid rockets coincided more or less with the availability of refined petroleum, since almost any liquefied petroleum product gives a greater yield of kilocalories per gram when burned compared to gunpowder. Because of this, liquid-fueled rockets were developed nearly simultaneously in Germany, the Soviet Union, and the United States by scientist-engineers who for the most part were unaware of one another's technical work.

As with any spacecraft, *Galactica* is dependent upon a source of fuel, and one basic fact of rockets has always remained constant—they use expanding gases escaping in one direction to provide thrust in the other direction in a direct application of Newton's Third Law. If you shoot a gun, you experience an excellent example of this law: the bullet or shot propelled from the barrel causes the gun to recoil. It kicks in the opposite direction. Were you standing on a skateboard when you fired the gun, you would roll in the opposite direction—the gun would provide thrust. In the case of a spacecraft, replace "shot from a gun barrel" with "hot gasses from a nozzle," and you have the basic concept of a rocket.

Two metrics can be used to gauge the performance of a propulsion system: thrust and specific impulse. Thrust is an instantaneous measurement of how much force is being generated by the propulsion method. More thrust means higher acceleration. High thrust is necessary, among other things, to propel a spacecraft off a planet and into space. Scientists and engineers also measure the sum total amount of rocket propulsion available in a given amount of mass of a fuel source, its efficiency, by using a value known as specific impulse (SI or I_{sp}). Note that there is no time dependence implied in the definition of SI. SI is the measure of the maximum total change in momentum that a propulsion system, or propellant, will yield per kilogram of fuel. This can occur over a very short or very long period of time.

With that understanding, let's take a look at some of the various propellants *Galactica* and her Vipers could potentially use, keeping in mind their missions. If *Galactica* is the Colonial equivalent of an aircraft carrier, then it likely has to get on station in a relative hurry. We know from the Miniseries that her FTL engines have not been used in over 20 years, so that implies that *Galactica*'s sublight engines allow her to get up to a reasonably high speed—enough to traverse the interplanetary distances within the Twelve Colonies in a reasonable amount of time. This would argue in favor of a propulsion system that can generate a high thrust. Vipers, on the other hand, have to intercept inbound threats in a hurry, another argument for high thrust (assuming Vipers and *Galactica* use the same kind of fuel). In both cases, the ability to stay on station for a long period is a benefit, which means that the propulsion system should be one with a high efficiency, or high SI, as well.

In the table on page 202 we compare the specific impulse and thrust of one type of solid propellant with several liquid propellants. Thrust

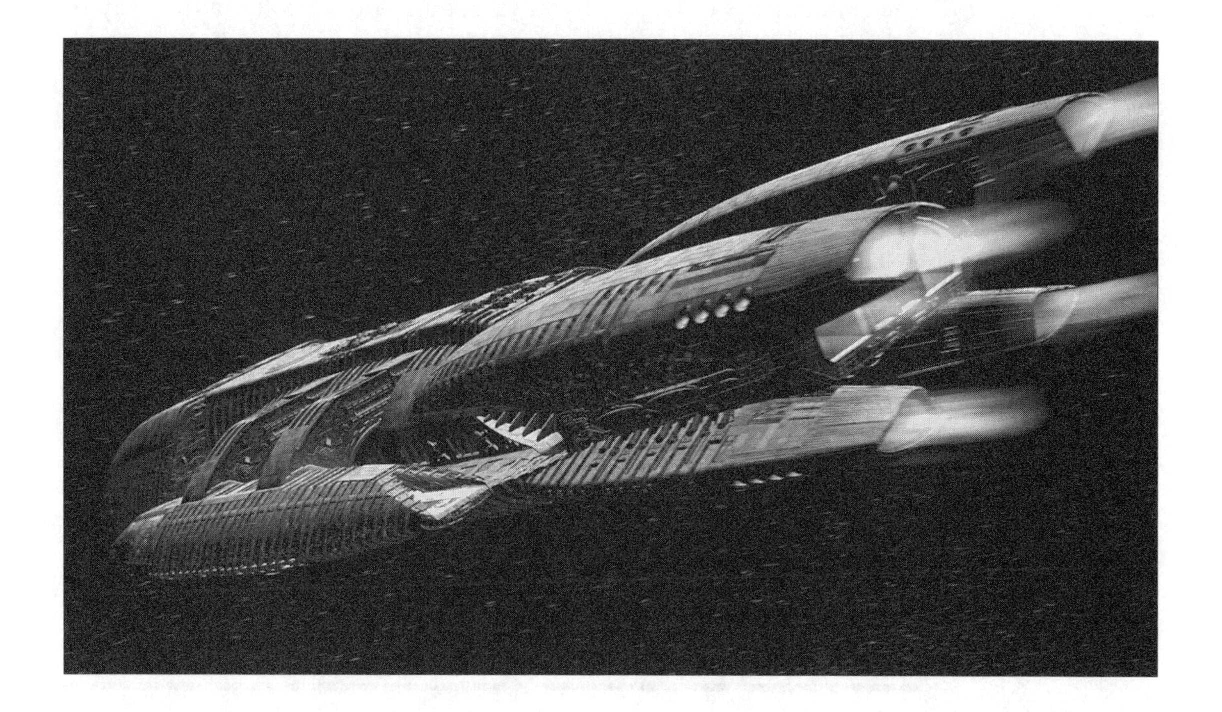

***Galactica's* sub-light propulsion system.**

in particular can be affected by many factors beyond the chemical properties of the propellant (including flow rates and the geometry of the rocket nozzle and combustion chamber), so the thrust values given should be viewed as an "all things being equal" comparison.

Nearly all these fuels present some difficulty. Liquid fluorine, for example, is nasty stuff. It's highly reactive with nearly any other element, which is a good thing for a rocket fuel to be. But it is also highly reactive with nearly any other element, which means it is amazingly difficult to store because it will react with its container. Hydrazine and tetrafluorohydrazine, both fluorine-based, are also extremely toxic and difficult to store. Kerosene and methane are easier to handle, but they're not as powerful, and kerosene requires either a source of raw petroleum or a pretty elaborate chemical plant. And beryllium infusion requires a constant supply of beryllium—not impossible to manage, but why bother? Beryllium is an uncommon element.

It turns out that the best all-around rocket fuel, in terms of ease of supply, ease of storage, and specific impulse, is also the simplest: liquid hydrogen and liquid oxygen. Since hydrogen and oxygen are the components of water, if you've got a source of water and a way to split the water molecules, you've got a source of rocket fuel.

Unfortunately, liquid oxygen and liquid hydrogen are among the most bulky kind of fuel a spacecraft can use. *Galactica* needs a much

Propulsion Type	Specific Impulse	Duration
APCP Solid propellant	~250	Seconds to minutes
Kerosene—liquid oxygen	~350	Seconds to minutes
Methane—liquid oxygen	~360	Seconds to minutes
Hydrazine—tetrafluorohydrazine	~380	Seconds to minutes
Liquid hydrogen(LH2)—liquid oxygen(LOX)	~450	Seconds to minutes
Liquid hydrogen(LH2)—liquid fluorine(LF2)	~460	Seconds to minutes
Liquid hydrogen/beryllium—liquid oxygen	~530	Seconds to minutes

more efficient type of propellant. It's time to look past liquid fuels to more exotic forms of space propulsion.

At this point it is important to recall that thrust and specific impulse should not be confused with each other. Thrust is a measure of the instantaneous force generated by a system; specific impulse is a time-independent measure of integrated thrust per unit of propellant that is expended. Propulsion systems with very low thrust can have very high specific impulses. In fact, in many cases, for most of the propulsion systems listed the two values seem inversely proportional.

Nuclear thermal engines, though included in the "exotic" list, have actually been built here on Earth. In fact, they were going to be the keystone of NASA's plan to land humans on Mars by 1980. They're remarkably simple in design: a nuclear reactor heats hydrogen gas to enormously high temperatures, then shoots the hot gas out the back of the rocket. In reality, building such a rocket that didn't blow up or contaminate the environment was next to impossible. NASA's nuclear thermal development project was shut down in 1972 when it became obvious that we weren't going to Mars by the end of the decade.

Propulsion Type	**Specific Impulse (seconds, typical)**	**Thrust (pounds, typical)**	**Duration**
Solid propellant APCP	250	2.65×10^6	Seconds to minutes
Chemical propellant	350–530	3.4×10^7 (Saturn V)	Seconds to minutes
Nuclear thermal	900	8×10^5	Seconds to minutes
Arcjet	1,600	< 10	Minutes
Pulsed inductive thrusters	5,000	< 100	Months
Electrostatic ion thrusters	8,000	< 5	Months to years
Variable specific impulse magnetoplasma rocket	30,000	0.25 pounds per 100 kilowatts	Months
External pulsed plasma propulsion	100,000	$> 10^8$	Years
Matter-antimatter	200,000	$> 10^9$	Years

Arcjet rocket engines run a stream of propellant (such as hydrazine or ammonia) past an open electrical discharge (sort of like a continuous electric spark). This energizes the propellant and makes it exit the rocket engine more quickly. The down side to this design is that it requires additional equipment to generate the electricity necessary to produce the continuous spark, for only a fourfold increase in specific impulse over conventional liquid propellants.

A pulsed inductive thruster is a form of ion propulsion that uses perpendicular electrical and magnetic fields to propel ionized gas into space. Like regular ion thrusters, it produces very low thrust that can be kept up for a very long time. Unlike regular ion thrusters, PITs can be scaled up relatively easily by increasing the number of pulses per second (which means increasing the energy available to the electrical and magnetic fields). To propel something the size of *Galactica* might require the electrical power used by a city.

Ion engines were first proposed by Robert Goddard back in 1906, and they have been a stalwart in science fiction for over 50 years. Electrostatic ion thrusters work like an old-fashioned television set. Neutral propellant gas, usually xenon, is injected into a discharge chamber (imagine the picture tube of an old TV). A cathode ray tube (the cylindrical hump at the back of the picture tube) sprays electrons into the chamber, turning the neutral xenon into positively charged xenon ions. Electric or magnetic fields then accelerate a beam of these ions out the back of the spacecraft, generating thrust. In this sense, an ion engine can be considered a mini-linear accelerator. Before they leave the spacecraft completely, the ions pass negatively charged electrical grids that reattach electrons to the ions. This prevents the spacecraft from becoming electrically charged. Like all ion engines, this design trades low thrust—less than the weight of a sheet of paper here on Earth—for a very long operating life. For this reason, they are used primarily for station-keeping on Earth-orbiting satellites. Because they can run for weeks or months, ion engines have very high SI. They have recently been used as the main propulsion system for NASA's *Deep Space 1* and *Dawn* spacecraft, as well as the European Space Agency's lunar-orbiting *SMART-1*.

A variable specific impulse magnetoplasma rocket (VASIMR) works much like the previously mentioned ion engines, with one huge difference: it can vary its output to provide low thrust/long life propulsion, or high thrust/short life propulsion. Theoretically, a VASIMR could be throttled up to take off from the surface of an airless moon, then throttle back to provide slow, long-term acceleration in outer space. While these might not be the engines *Galactica* uses, a version that somehow works in an atmosphere would be perfect for Vipers and Raptors.

External pulsed plasma propulsion used to go by the older name of "nuclear-pulsed propulsion." It was a plan to explode small nuclear bombs behind (or even within) a spacecraft. The radiation from the explosion would vaporize a portion of a large "pusher plate" mounted on the back of the spacecraft. The vaporized material that shot off in one direction would propel the spacecraft in the opposite direction. By exploding a bomb every few seconds, a massive spaceship (early designs were the size of a battleship) could putt-putt its way to Mars in a few weeks, as opposed to the more typical 6 to 12 months, leaving a trail of deadly ionizing radiation behind it.

Recall that our battlestars and Vipers require a propulsion system with a combination of high thrust and high SI. Of the methods we've discussed so far, VASIMR or a variant is the only candidate for that, and not a particularly good one. Matter-antimatter engines are an old *Star Trek* standby that can probably be ruled out. The ability to produce and contain antimatter seems to be beyond the technology of the Colonial Fleet.

Enter tyllium.

For decades, science fiction writers and authors have invented new and exotic materials to power their spacecraft, or to endow their spacecraft/armor with the desired combinations of weight, strength, and other material properties. Borrowed from real-life engineers, a name has even been coined for fictitious substances that have such improbable combinations of material properties: unobtanium.*

Unobtanium in the *Star Trek* universe is dilithium; in *Battlestar Galactica* it is tyllium. Tyllium was the fuel source for *Galactica* and

*And yes, in *Avatar* James Cameron was having his little joke by naming Pandora's magic mineral "unobtanium."

FINDING MATERIALS TO MAKE THE BLACKBIRD

When the writers David Weddle and Bradley Thompson wrote "Flight of the Phoenix," the executive producer Ron Moore insisted that the materials used to make this craft shouldn't simply appear out of nowhere, as might be the case on other television dramas. The Blackbird, he said, should be made from materials the viewer can expect to exist within the Fleet.

In a Fleet depleted of resources, it is amazing what one can come up with when looking hard enough. After the uprising and subsequent pardon of the prisoners on the *Astral Queen*, it's reasonable that there might be a lot of iron—in the form of jail bars—available for

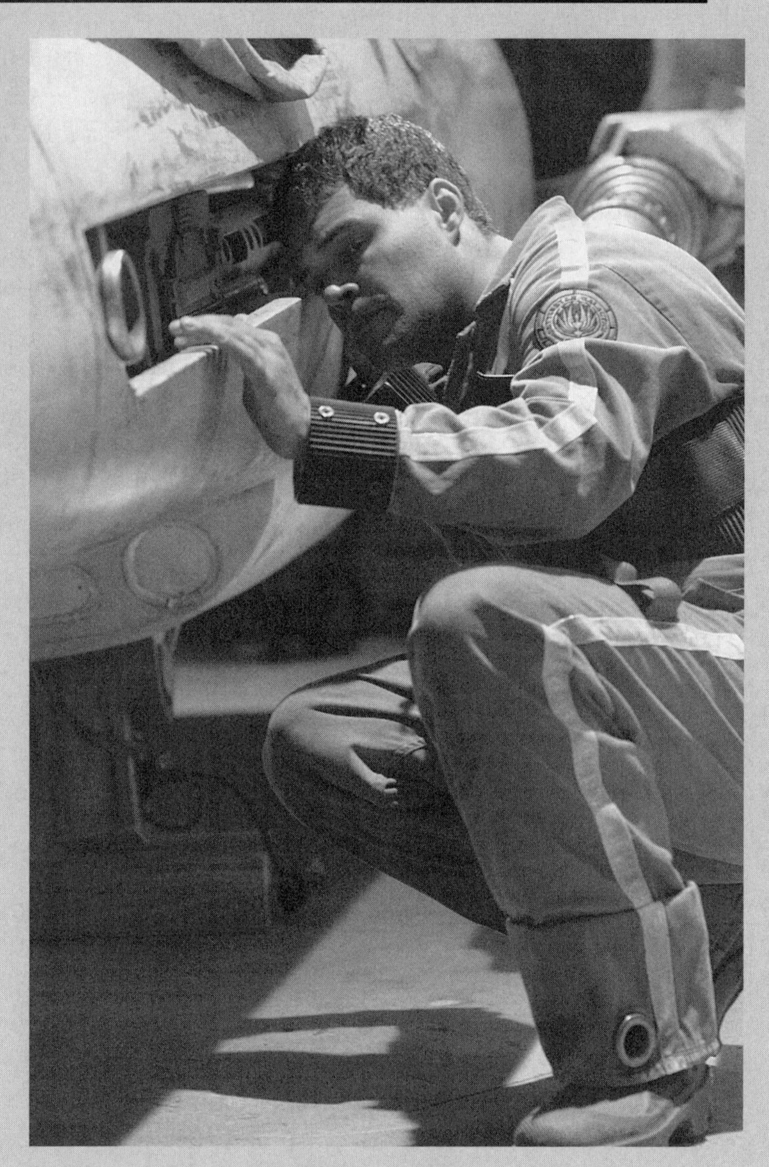

Chief Galen Tyrol in "Flight of the Phoenix."

›››

››› an airframe. Battle damage might have created some usable rubbish—in particular, sheet metal that could be used on the aircraft. We know that the engines were spares taking up space on the flight deck of *Baah Pakal*.

Where on Caprica did the carbon composites come from for the airframe? Recall that the huge luxury liner *Cloud Nine* had a lake (before *Cloud Nine* was destroyed by the explosion of a nuclear warhead in "Lay Down Your Burdens, Part II.") The resin used to repair paddle boats provided the outer skin for the Blackbird!‹

her Vipers (and presumably Raptors as well) in both the original series and the reimagined series, with a minor difference in pronunciation. (In the original series *tyllium* was pronounced TIE-lee-um; more recently it has become TILL-ee-um.)

We have no way to perform a direct comparison of the thrust generated by a tyllium-powered propulsion system to that of modern-day solid, chemical, or even exotic propulsion systems. We also do not have a way to compare relative values of specific impulse. What we do know, however, is the *enthalpy* of tyllium.

Propellant	Enthalpy (joules per kilogram)
APCP	3.1×10^7
LH2/LOX	1.3×10^7
Methane/LOX	1.1×10^8

In thermodynamics, the enthalpy of a chemical reaction is a measure of its thermodynamic potential—how much energy is locked up within a given mass of chemical reactants. While the total enthalpy is a difficult quantity to measure, the enthalpy change of a chemical reaction is useful and easier to measure, since it is a measure of the potential work that the reactants can perform. For example the APCP in the space shuttle yields 31 million joules of energy for every kilogram burned, or 3.1×10^7 J/kg.

In comparison, the exotic propulsion system that best optimizes thrust and specific impulse is VASIMR, which can generate 4.3×10^{11} joules per kilogram of propellant.

In the episode "The Hand of God," Dr. Baltar says that the enthalpy of tyllium is "half a million gigajoules per kilogram," or 5×10^{14} J/kg. If 1 kg of tyllium is converted directly into energy by $E = mc^2$, that

THE BLACKBIRD'S DDG-62 ENGINES

If an author wishes to write convincingly on a topic, he or she has to do a boatload of research. The willingness to do so can be the difference between a good writer and a not-so-good one.

My friendship with Brad Smith dates back to seventh grade. During the first two years of the run of *Battlestar Galactica*—until his career took him to bigger and better things—Brad, known by his crew as Captain Smith, was the commanding officer of the *Arleigh Burke* class guided missile destroyer USS *Fitzgerald* (DDG-62).

Each type of ship in the U.S. Navy has a given designation. An aircraft carrier is a CV, a cruiser is a CG, and a destroyer is a DD. If the destroyer has the capability to launch guided missiles, then it is referred to as a DDG. The hull number of *Fitzgerald* is 62, hence the proper designation of USS *Fitzgerald* is DDG-62.

Before moving to Japan, the "Fightin' Fitz" had a Tiger Cruise just off the coast of San Diego. In Navy-speak a Tiger is any relative or friend of a ship's crew member, so a Tiger Cruise is essentially "Friends and Family Day." The ship put to sea for a few hours, and the visitors got to see first-hand what life is like aboard a U.S. Navy warship.

Since much of *Battlestar Galactica* is, in fact, about life aboard a warship—albeit one in space—I figured that this presented a great opportunity for our writers. With the captain's permission, I invited the entire writing staff of *Galactica* to join us on *Fitzgerald* that day. Many were interested, some had previous commitments, a couple more had to cancel at the last minute. Ultimately only Bradley Thompson and his significant other, Peggy Sue, were able to attend.

While on *Fitzgerald* Bradley saw and experienced as much as possible, and made the most of his time aboard. To him this was research, so he took copious notes during his day-long visit. Although it might be a stretch to say they had a direct influence, those notes and that experience subsequently impacted the writing of the first two episodes of season two, "Scattered" and "Valley of Darkness." Bradley's experience had a very small, but more easily observable, influence on two other episodes, "Flight of the Phoenix" and "Pegasus."

Recall that in the episode "Flight of the Phoenix," Chief Tyrol made the decision to build the Blackbird stealth craft. It looked like construction would come to an end, sadly, because there were no engines available. Commander Adama supported the construction of the Blackbird, but insisted that no parts be used that could have value in operational Vipers or Raptors. A solution came from the most unlikely of places: Colonel Tigh. Claiming that he owed the XO of the *Baah Pakal* a favor, Tigh said that there was a pair of "old DDG-62 engines

›››

taking up space in their cargo hold." The Blackbird was the recipient. In the next episode, "Pegasus," aeronautical engineer Peter Laird took note of the DDG-62 engines while inspecting the Blackbird, and commented to Tyrol that he had designed them.

The DDG-62 reference was a thank you to the captain and crew of USS *Fitzgerald* for their hospitality.

would release 9×10^{16} J/kg. If Dr. Baltar is correct, the energy locked away within tyllium is slightly over half a percent of the yield of a direct conversion of mass to energy (which is equal to the output of a matter-antimatter reaction). Put differently, there is several million times more energy locked away in tyllium than within the most powerful chemical rocket fuels known today. Tyllium even has thousands of times more potential energy than the most promising exotic propulsion technologies on the near-term horizon. Consider the places humankind might have ventured today with access to tyllium-based propulsion!

From appearances, the blue exhaust from *Galactica*'s engines looks similar to the exhaust of ion or magnetoplasma engines, so that's probably how they travel within the Twelve Colonies. Perhaps tyllium, then, is used to generate electrical power, which, in turn, powers a more exotic type of engine.

To travel between star systems, they use another form of propulsion entirely.

CHAPTER 22

Faster Than Light: *Galactica*'s Jump Drive

Most of space is empty and there aren't a lot of strange things to bump into.
—Ronald D. Moore, *Battlestar Galactica* series bible

It is a fundamental mantra of the screenwriter: never wake the audience from your dream. A television concept can be the most poignant or beautiful work imaginable, but careless writing can instantly turn an "Oh wow!" moment into an "Oh *please*!" moment. Works of science/speculative fiction like *Battlestar Galactica* tend to attract a more technically literate fan base, which means that egregious technical gaffes are more readily spotted and denigrated ("Worst! Episode! Ever!"). There is an implied contract between those who create and those who enjoy science fiction: In defining the laws under which the artist's fictional universe operates, the artist is allowed only a few "conceits"—ways in which the laws of the known universe may be bent or broken. If the artist subsequently remains faithful to the laws he or she has defined, we the audience will collectively and happily suspend our disbelief and allow ourselves to be taken on an adventure. But the elastic of the audience's collective psyche will stretch only so far. It's best if even the conceits have reasonable explanations.

> **Space is big. Really big. You just won't believe how vastly hugely mind-bogglingly big it is.**
> —Douglas Adams, *The Hitchhiker's Guide to the Galaxy*

Like artificial gravity, faster-than-light travel (aka FTL) has been one such conceit of science fiction since the inception of the genre, and for good reason. Although much of the appeal of *Battlestar Galactica* is the human drama, the political intrigue, and the never-ending specter that the next Cylon attack could be the final Cylon attack, it is beneficial if the Rag Tag Fleet actually has a chance of getting to Earth within the lifetimes of our central characters or, preferably, over the run of the series. That means that the Fleet has to cover the vast distances between stars in a short amount of time. The problem is the Second Law of *The Science of Battlestar Galactica*: "Space is mostly empty. That's why it's called 'space.'"

A typical robotic Mars probe takes 6 months to a year to make its journey. It took the Cassini spacecraft 6 years and 8 months to reach Saturn. The most distant object ever built by humankind at present is the *Voyager 1* spacecraft. In early 2010, *Voyager 1* was slightly over 112 astronomical units (AU) from the Sun, and it took more than thirty years to get there. *Voyager 1* is currently traveling into interstellar space at 3.6 AU per year in the general direction of the constellation Ophiuchus. The spacecraft will eventually pass within 1.64 light-years of the small red dwarf star AC +79 3888, currently located in the constellation Ursa Minor, in approximately 40,000 years. So not only is the distance to even the nearest star, well, astronomical, but current technology has not yielded a propulsion system that can cover interstellar distances within a human lifetime.

Is interstellar travel even possible, when distances are measured in light-years as opposed to AU? As Admiral Adama once said, "Context matters." Let's put into context, then, the scale of how vast and empty our galaxy truly is. Let's say that Sol, our Sun, was an orange sitting at the end of a basketball court in New York's Central Park. In that scale, Earth would be the size of a grain of dust 70 feet away, about the

Laura Roslin and William Adama.

Saul Tigh and Laura Roslin.

distance from the baseline to the top of the opposing key. The nearest star to Sol is the trinary star system Alpha Centauri, 4.37 light years away, which would be represented by a slightly larger orange (Alpha Centauri A), in mutual orbit with a large plum (Alpha Centauri B), with a ball bearing in orbit about the pair (Proxima Centauri). At that scale, all three stars would be 3,663 miles (5,895 km) away, just east of Paris, France!

Centaurus is a constellation in the southern sky, and since the bulk of Earth's population is in the northern hemisphere, Alpha Centauri is below the horizon for most people. The closest star visible to nearly all of Earth's population is also the brightest star in our sky: Sirius, 8.6 light years away. Relative to our orange in Central Park, Sirius would be a grapefruit resting outside of Karachi, Pakistan, 7,210 miles (11,604 km) away. If *Voyager 1* were headed toward Alpha Centauri, it would take between 75,000 and 80,000 years to get there; if it were headed toward Sirius, it would take between 148,000 and 157,000 years.

Clearly, if our brave men and women aboard *Galactica* are ever to make planetfall again in their lifetimes, we're going to have to pick up the pace a bit. We know we can't just slap a gargantuan rocket engine and huge fuel tanks onto *Galactica* and gain more thrust that way. In chapter 11, "Special Relativity," we examined how parameters like the length of a spaceship and the passage of time on board are velocity-dependent. It turns out that Special Relativity has another nasty trick up its sleeve: as an object accelerates to relativistic speeds, it behaves as if it has more and more mass. The faster your spaceship travels, the harder it is to increase its speed still further. You get far less bang for the buck out of your fuel supply

What about forms of propulsion like nuclear propulsion? Antimatter pulse-detonation propulsion? Exotic propulsion systems like these may get a spacecraft up to relativistic speeds, in which case our crew will have the opportunity to visit, perhaps, a handful of star systems in a lifetime. So from a dramatic standpoint, we're going to have to travel at faster-than-light speeds, and there's no two ways around it. How might we plausibly do that in an "Oh wow!," not an "Oh *please*!," fashion?

Nothing we manufacture can travel at the speed of light in a vacuum. But here's a funny little trick: although an object can never travel at exactly the speed of light, the equations of Special Relativity do not prevent an object from moving faster than the speed of light. If there is a way to travel from our slower-than-light realm into the faster-than-light realm without ever actually traveling at exactly the speed of light at some point—a way to "tunnel" through light speed—perhaps superluminal speeds are attainable.

Tachyons are theoretical subatomic particles that live in the FTL realm; their lower speed limit is the speed of light, and it is impossible for them to travel any slower. As you might guess, this quality has made tachyons a staple of science fiction ever since they were first described in 1967. In real life, the unfortunate thing about tachyons is their annoying habit of disappearing before their detection, since the equations of Special Relativity also say that anything traveling faster than light would also travel backward in time. Therefore, any reasonable description of the function of *Galactica*'s FTL drive would have to account for, or ignore, Special Relativity. Since the show uses a relativistic effect as a plot device, it would be highly inconsistent, not to mention a bit hypocritical, to use or discard relativity at a "dramatic whim."

In his online blog posting of January 30, 2005, the *Battlestar Galactica* executive producer Ronald D. Moore gave a few hints about *Galactica*'s FTL drive:

> An FTL Jump is nearly instantaneous, essentially moving a ship from point A to point B without traveling through the normal space-time continuum, presumably by bending space around the ship in some way. The analogy I used during production was to imagine three dimensional space as a flat piece of two dimensional paper. To get from one side to the other, you can travel in a straight line across the page, or you can gently bend the sheet in half and cross from edge to edge virtually instantly. How this is accomplished and what is the basis of this technology outstrips my technical brainpower.
>
> In fact, I feel faint just coming up with that explanation.

In the series bible, Ron also provided another constraint:

> The speed of light is a law and there will be no moving violations.

So, in summary, we have to describe a system that will transport the mass equivalent of a small asteroid across the cosmos, from point A to point B, without traversing the space in between. We must have functional speeds faster than light while never actually breaking the speed of light in the process. Hence the "jump drive."

In *Galactica*'s case, we are looking at flinging a quantity of mass on the order of 120 teragrams (120 billion kilograms, or about 132 million tons) across the cosmos. Is that even remotely possible, to send something approaching the mass of a thousand *Nimitz* class aircraft carriers several light-years away? Or is this where the First Law of BSG comes into play? Obviously, no human alive knows exactly how an FTL drive works. Nevertheless, we can ponder some of the relevant physics that may one day lead us to FTL travel, and will yield a drama-driven plot element based upon known physics. The constraints are now in place.

Teleportation

Can we teleport the mass? The idea of a *Star Trek*–like transporter is to convert matter to energy, send that energy to a specific place, and then reconvert that energy back into matter. If *Galactica*'s FTL drive works like a *Star Trek* transporter, its problem is to convert 120,000,000 kgs of battlestar completely into energy. We've already established that there is an enormous amount of energy stored in even the smallest amount of matter; with *Galactica*'s mass, we're talking a significant amount of juice.

Yes, that's a problem.

In the first atomic explosion in the New Mexico desert in 1945, the bomb held a little more than 6 kilograms of plutonium. When the fallout and ground contamination from the explosion were examined, Manhattan Project scientists estimated that perhaps 1 kilogram of that plutonium had actually been involved in nuclear fission, and that almost precisely 1g of mass had been transformed into energy. If the purpose of the FTL drive is to convert the entire mass of *Galactica*—all 120 trillion grams of it—to energy instantaneously, the release of energy would be the largest explosion seen in this part of the galaxy since the supernova that ultimately led to the creation of our Sun. Explosions like this, at varying magnitudes, would happen every time any ship in the Fleet jumped anywhere. Under the category of "potential jump drive physics," we can rule this out.

Hyperspace

What about hyperspace? A common device used in science fiction and science fantasy depends upon the existence of an alternate realm where space is "denser" and the speed of light is not a speed limit, or perhaps there is no speed limit at all—hyperspace. A spacecraft enters hyperspace through its own or by external means (by using a jump gate versus a jump point in *Babylon 5* terminology), it travels rapidly to the destination, and reenters "normal" space. Such a dimension has been called hyperspace (*Babylon 5, Star Wars*, and others), subspace (*Star Trek*, though the plot device is used for superluminal communications only), or slipspace (*Star Trek: Voyager, Andromeda, Doctor Who*, and the *Halo* series of video games). When the concept of hyperspatial travel initially appeared in 1930s science fiction, there was no corresponding science to explain it. It was a dramatic conceit. Comparatively recent models of the structure of our universe may provide a dramatically satisfying post hoc explanation for hyperspace.

The notion that there may be parallel universes coexisting simultaneously has been a frequently used device in science fiction, though the concept of multiple universes and/or multiple realities goes back much further and can even be found in ancient Hindu writings. Parallel universes have been generally depicted two different ways in science fiction. In one case, a parallel universe is a universe that is almost exactly like ours, but where some small deviations in history have propagated to create dramatic differences. In other cases a parallel universe is a distinct universe that exists adjoining to our own—like E Space and N Space in *Doctor Who*, or fluidic space in *Star Trek: Voyager*. Hyperspace relies on the second type.

Branes in the multiverse.

Recent advances in superstring theory, specifically a concept called brane cosmology, hint at the existence of hyperspace. The fundamental premise of brane cosmology is that our four-dimensional space-time may be one of many (normally) disconnected universes separated by some sort of membrane, or "brane." In other words, our four-dimensional universe is one of many that coexist simultaneously within a five-dimensional space known as the Bulk. Scientists have proposed cosmological brane models that suggest that interactions with neighboring branes may explain the weakness of gravity (compared to other fundamental forces). One scenario, called the ekpyrotic universe, is based upon the hypothesis that our observable universe came into being when two branes collided. Together, all the parallel universes along with the Bulk form what has been called the multiverse.

The reason the Bulk has proved seductively attractive to SF writers is that there is no guarantee that the physical laws that prevail in our universe are the same in the Bulk between universes. Want to travel beyond the speed of light without the hassles of Special Relativity? It just may be possible in the Bulk.

Think of the multiverse as a huge apartment/condo/co-op building. Our own universe, and every other universe, would be like individual apartments in that larger structure. Suppose we live in universe 4D and we want to go from the bedroom to the kitchen. The house physics rules for our particular unit say that we can't travel any faster than the speed of light within our universe. But there's nothing saying that we can't go outside the apartment—into the empty hallway between the universes, where the speed limit would not apply—and get to the kitchen that way. The Bulk—the empty hallway between universes—just represents a region through which we may be able to travel. It fits the concept of hyperspace.

Of course, there are no guarantees that the rules of the Bulk would allow faster-than-light travel. But we do now suspect that there is a realm that has some of the properties that we attribute to hyperspace—outside our own universe, difficult (or impossible) to access, with different rules and limits. For a science fiction writer, taking a plausible but hazy scientific concept and giving it the attributes you need for your story is vastly preferable to just making up something like "hyperspace."

Wormholes

Like hyperspace, another science fiction FTL concept is the use of wormholes. Wormholes are shortcuts through space-time that allow travel from point A to point B at a functional speed greater than *c*. The object traveling through the wormhole never exceeds *c*; it just has such a shortened path between the two points that it's as if it traveled faster than light in normal space. Unlike hyperspace, which started out solely as a child of science fiction, the concept of wormholes arises from honest-to-goodness real mathematical solutions to Einstein's General Relativity equations.

Although the concept of shortcuts through space-time dates back to 1921, the term "wormhole" was first coined in 1957 by the physicist John Wheeler—using the analogy that a worm chewing through an apple between two points travels a shorter distance than a worm crawling across its surface. Although several types of wormholes have been mathematically discovered over the years, it was not until 1988 that researchers described wormholes that could be both stable enough and large enough for a spacecraft to travel into. Since then, wormholes have been discovered that not only connect distant points within this

The geometry of a wormhole.

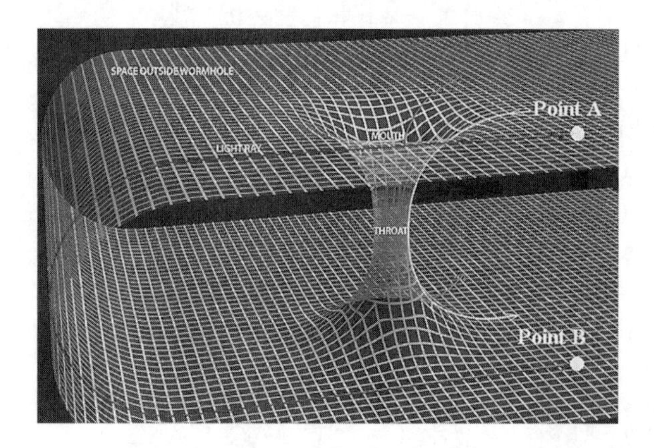

universe, but points within adjoining universes. Perhaps inter-brane wormhole travel would make a great future science fiction series!

While wormholes and related cosmological structures are excellent fodder for science fiction FTL travel, they don't quite explain the operation of *Galactica*'s FTL drive. Travel through a wormhole can be functionally FTL, but as has been depicted in series like *Star Trek: Deep Space Nine* and *Stargate: SG-1*, there is still a certain amount of travel time involved, and this is not what we've seen on *Battlestar Galactica*.

Space Warp

Both theory and observations of General Relativity make it clear that space-time can be curved, warped, perhaps even folded or compressed. One of the more famous FTL plot devices in science fiction is *Star Trek*'s warp drive, a propulsion system that continuously compresses, or warps, space ahead of a starship. It was one of those ideas that was "in the air" in 1960s science fiction. Just prior to the original *Star Trek* series, in his 1965 novel *Dune*, the author Frank Herbert wrote that the Guild Heighliners moved at superluminal speeds by using a phenomenon called the Holtzmann Effect to travel through foldspace, a similar concept.

Imagine an ant crawling on a blanket. If the ant could somehow scrunch up the blanket in front of it and "unscrunch" the blanket behind it, it could make its way across the blanket much more quickly than if the blanket were flat. The USS *Enterprise* and the Heighliners work the same way. In both cases the spacecraft do not actually exceed the speed of light locally; they merely create denser regions of space through which they can pass and effectively travel superluminally.

In a strange twist of "science imitates art," nearly 30 years after *Dune* and the original *Star Trek*, the physicist Miguel Alcubierre described a mechanism by which a spacecraft could travel within a "warp bubble," creating a compression of space-time ahead of the craft and a spreading-out of it behind. Although there are serious technical hurdles that confine the Alcubierre Drive to the realm of the theoretical—chief of which is the paradox that you need to have a functional Alcubierre Drive before you can create one—this is the direction we have pursued in explaining the function of *Galactica*'s jump drive.

Actually . . .

One of us was tasked with determining how the jump drive on *Galactica* works for the season two episode "The Captain's Hand." Page 221 is a transcription of the technical notes submitted to the production staff.

Most similar to the Heighliners from *Dune, Galactica* generates a field that warps/folds space ahead of it, not just locally like a warp drive, but over the trajectory of the entire jump. It passes the warp over itself, "attaches" itself to the other side of the fold, and switches off the field. Voilà, a space-time jump! Keep in mind that this is an explanation within the world of science fiction, and is never intended to be practically attainable. Nevertheless, work is under way on this very front.

From 1996 to 2002, NASA sponsored a program at the Glenn Spaceflight Center near Cleveland called the Breakthrough Propulsion Physics Program. This was an effort designed to investigate new areas of physics that could literally lead to quantum leaps in our ability to travel the cosmos rapidly. The goal of the program was not to develop technology for traveling faster, but rather to explore and understand the physical principles that can be exploited by future technology to make significant advances in propulsion. Although the program found no breakthroughs to take us to the stars soon, it did generate sixteen peer-reviewed articles that may shed light on future research directions. Perhaps, 30 years hence, a brilliant young physicist will find a theoretical basis for the jump drive the way Alcubierre has done for warp drive. If we're particularly fortunate, maybe some clever engineer will find a practical way to implement the Holy Grail of space flight and literally open the galaxy to human exploration. That would be the ultimate "Oh, WOW!" moment.

The Captain's Hand

BLUE Production Draft
TECH Calculations/Comments/Observations
Kevin R. Grazier
5 October 2005
General Comments on FTL Travel and the FTL Drive

How does an FTL jump work? We've established that the Colonials are far ahead of present-day Earth in some technologies, but because they have turned their back in some cases (computers), fairly equal in other technologies.

Let us assume that Colonial science has done one thing we've not: they've unified the electromagnetic force with gravity—in other words the two forces are, at some level, different manifestations of one single force. The Colonials understand how and why.

If scientific understanding of this has led to technological application, then perhaps the generation of an extremely intense electromagnetic field can be the equivalent of an intense gravitational field. One implication of general relativity is that the presence of matter bends/warps space. That's what gravity is—warped space. So perhaps an FTL drive creates an intense EM/gravimetric field that either creates a region of dramatically warped space ahead of a vessel—or a rip in the fabric of spacetime which the ship uses to enter into hyperspace.

I'll think about this more later, right now we have a show to get out . . .

So to do this we would need a field generator. The intense EM/gravimetric field needed for FTL travel would likely do awful things to biological systems (that is, people) who were exposed to it, unfortunately. So let's assume that an FTL drive consists of multiple field generators—that way no single generator creates a biologically harmful field. Through the process of constructive/destructive interference, the generators are "tuned" to create a "jump point" directly in front of the ship. Then, either the main drive propels the ship into the jump point, or the field is modulated to "pull" the jump point across the ship—the latter description better describing the visual effects we've seen to date.

So let's assume that Galactica has at least four field generators (two starboard/two port; two up; two down). I assume that the energy required to create the initial warp in space is far greater than that necessary to maintain travel through it.

Also, if one generator goes offline, we don't jump (or, perhaps, we jump somewhere into BFA—bumfrak Aerilon).

I also assume that the "spinner" portion of the FTL drive refers to a power source that generates a colossal amount of power in a very short amount of time—a huge pulse to the field generators. The generation of such an intense energy field, and the generation of the power necessary to create such a field, is likely to generate a lot of heat as a by-product. Dissipating this heat would be the function of one of the major subsystems of a jump drive.

Also, as the field generators fire up, they would likely create local phenomena, which explains the warping effect we saw in the miniseries.

I also assume that at some level, the artificial gravity on board our ships is a different application of the same technology.

CAFFEINE: THE ORIGINAL FTL DRIVE FUEL

I do a lot of public speaking, mostly about astronomical education and EPO (Education and Public Outreach) for the Cassini-Huygens mission to Saturn. I sat in a daze in the Portland airport, having just given a description of Cassini's expected science return from Saturn to a conference of military personnel and contractors. Since I had some frequent-flyer miles, and since Portland was (relative to L.A.) in the Vancouver area, I decided that after my talk, I would continue on to the *BSG* set for a two-day set visit. (Although I could do all of my work from L.A., occasionally I was allowed to visit the set and see the filming.) So in order to get my affairs in order enough to afford the three-day absence from my day job, I had been up all night working. Hence the daze.

Since Portland International Airport has free Wi-Fi, I figured that I'd check my e-mail while waiting. Just as the gate agent announced that it was time to board, I read a semi-urgent email from Ryan, *Battlestar Galactica*'s writer's assistant, asking that I get in touch with Bradley Thompson as soon as possible. I sent a brief reply that I was about to board a plane, and that I'd phone when I could.

The instant we landed in Vancouver, I flipped open my cell phone and made the call. "Hey Bradley, you know that annoying guy on every flight ever? You know, the one who gets on his cell phone the instant you land and talks really loudly? For the first time in my life, it's ME. What's up?" Bradley said, "You knew this was going to happen eventually. David [Weddle] and I are doing the rewrite on [episode] 216. We need you to figure out how the FTL drive works, so we know what components it has, so we know what can get battle-damaged, so the captain of *Pegasus*, in a *Wrath of Khan*–like move in the heat of battle, can run down to engineering and save the ship . . . I get in at 10:30 tomorrow, that's how long you have."

Frakkin' wonderful. Did I mention that I'd been up all night?

So I slept for a scant few hours, got up at 0-dark-30, and made a huge pot of coffee. As Bradley alluded in his phone call, we had previously held brief discussions on the FTL drive—not so much about how it worked, but more in the form of "we're going to have to describe this eventually." So I already had my own rumblings of ideas about how the FTL drive worked, but had never put them into a coherent formulation. Luckily, and wholly coincidentally, my own musings were not dissimilar to those shared by Ron in his blog posting.

So I paced my hotel room, and drank coffee. I mused on the science and technical aspects of the FTL drive, and drank coffee. I wrote up tech notes for the FTL drive, as well as other aspects of the new script, peed a lot, and drank still more coffee. I hit SEND

›››

on that e-mail at 10:28, and managed to get to the set in time to see Billy Keikeya get shot.

This "description" of the inner workings of *Galactica*'s FTL drive was the product of 5 sleep-deprived hours of pacing, pondering, writing, caffeinating, and peeing in a Vancouver hotel room. Over the years, at science fiction conventions, the single most frequently asked question I get is, "How does *Galactica*'s FTL drive work?" I have always given the same reply: "As long as this could be a plot point in a later episode, I'm not sayin'." I fully realize that this has generally been interpreted as "How the FRAK do I know?" It's nice to be able to share the details now, as well as confirm that, yes, we did have a clue.❮

CHAPTER 23

Artificial Gravity

If artificial gravity could be invented, how might it work? Firing a rocket engine in a spacecraft will create a sensation of gravity, in the same way that accelerating in a fast car will push you against the back of your seat, and taking off in an airplane makes you feel as if your stomach has been left behind on the tarmac. This acceleration feels just like gravity, and from the perspective of the universe, it is gravity.

Constant acceleration seems like the easiest solution to the artificial gravity problem: just keep firing *Galactica*'s engines at a constant rate in the proper direction, and presto, the acceleration will keep people and objects glued to the deck as though they were standing on Caprica itself. Instant artificial gravity. Unfortunately, constant rocket acceleration is probably the worst method of creating artificial gravity.

Let's assume that Caprica is the same size as Earth, with the same gravitational field. In order to create Caprica-equivalent gravity on *Galactica*, the

Samuel Anders.

Samuel Anders.

engines would have to fire in a direction perpendicular to the deck in a way that would increase *Galactica*'s speed by 9.8 meters per second. Every second. In the first second, the entire ship would be moving at nearly 10 meters per second. At the next second, it would be moving at almost 20 meters per second. After one hour, *Galactica* would be traveling at 6 kilometers per second, in a direction perpendicular to the decks, and still accelerating. At the end of five weeks the ship will be traveling at 10 percent of the speed of light, still in a direction perpendicular to the decks, and it'll just keep going. At the end of a year, assuming the fuel holds out, *Galactica* will be traveling at a large fraction of the speed of light, and it will become increasingly difficult to go any faster. Your artificial gravity is over.

The next simplest way to produce a sensation of artificial gravity is by rotating the spacecraft, or at least a section of it. Objects in the spacecraft will move with the rotating section, and this motion will eventually cause them to "stick" against the outermost rim of the rotating section. The force of inertia, known more colloquially as "centrifugal force," keeps crew members pinned against the hull in the same way that the spinning carnival ride Vortex keeps victims—uh, carnival goers—pinned against the wall.

Spinning the spacecraft works well in movies and on TV, but in real life there is one problem with this type of centrifugal ersatz gravity: the Coriolis Effect. It's kind of complicated, but the basic idea is that if you're located on or in a rotating environment, you'll start to rotate, too (or you'll spend a phenomenal amount of energy to stop yourself from rotating). The Coriolis Effect causes hurricanes to rotate counterclockwise in the Northern Hemisphere and the clockwise in the Southern Hemisphere, and in a smaller space it would do the same thing to a person's inner ear fluid, which regulates the sense of balance. In a rotating space station your eyes might tell you that your body is standing still, but your inner ears will "know" that you're spinning around. If you're lucky, you'll only get dizzy.

Experiments on Soviet cosmonauts showed that Coriolis nausea caused by spinning the spacecraft can be eliminated if the rotation rate is kept low—which naturally means that the ship, or at least the radius of rotation, has to be correspondingly huge. Most people have little trouble tolerating anything up to 3 rotations per minute (RPM) without throwing up. Some people never make it past 3 RPM, while others can acclimatize to 6 RPM. Practically no one escapes illness at close to 10 RPM. In summary, if the ship rotates too slowly, a large fraction of the crew is vomiting from space sickness; if the ship rotates too rapidly, a large fraction of the crew is vomiting from motion sickness. When humans leave Earth to explore space en masse, invest in the company that makes puke bags.

Obviously, *Galactica* does not spin to simulate gravity, but a ship named *Zephyr* does. *Zephyr*, also known as the "Ring Ship," is one of the most prominent ships in the Rag Tag Fleet. *Zephyr* seems to be quite the oddity within the Fleet, apparently being the only ship that creates artificial gravity in this way. This could imply that *Zephyr* is very old, but if that's the case then her FTL drive

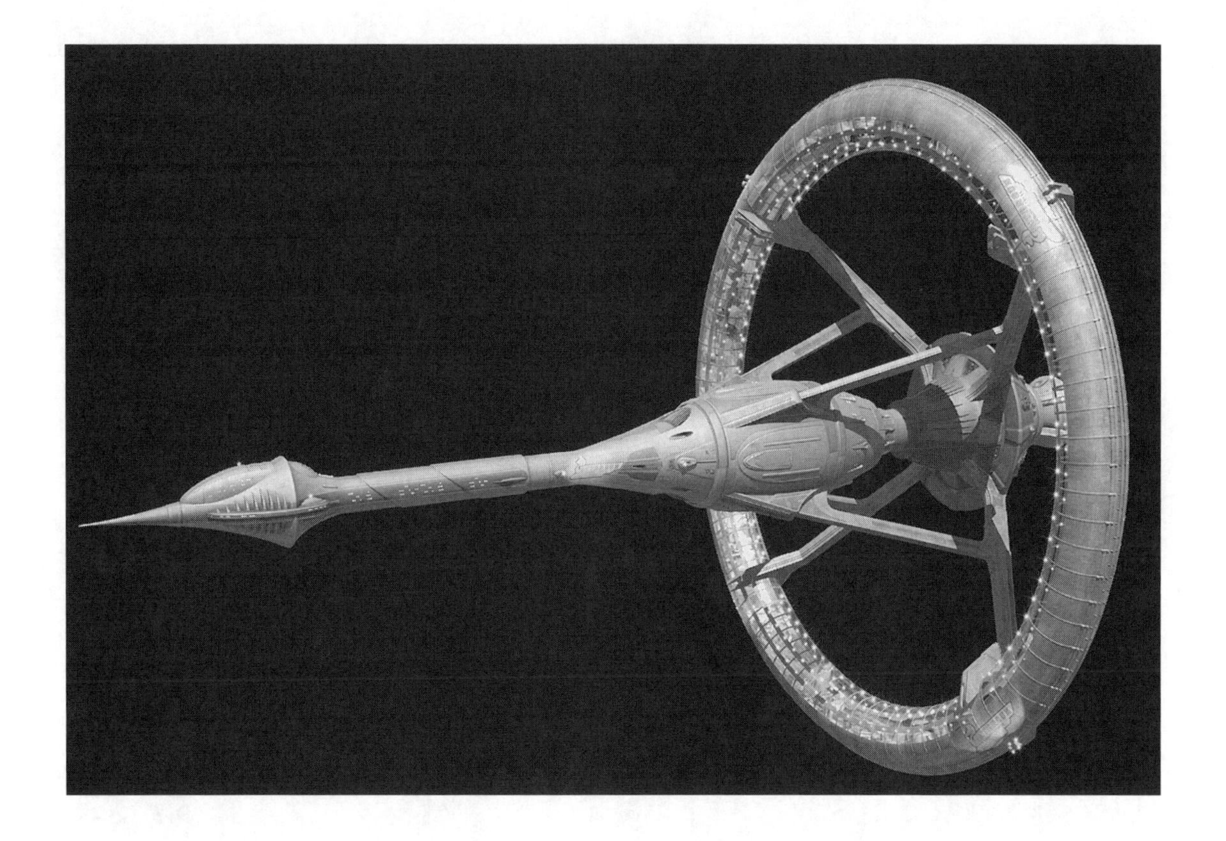

The "Ring Ship," *Zephyr*.

would have had to have been a retrofit. Perhaps, like a Chrysler PT Cruiser, the inner workings of the spacecraft are modern, but she has a "retro" exterior: "Come experience what space travel was like for our Kobolian ancestors!" *Zephyr* just may be a Colonial manifestation of steampunk.

Zephyr is one of the larger ships in the Rag Tag Fleet, and it stands out. Every viewer has noticed that gigantic ring slowly spinning. Is that realistic, though? Should it be spinning faster? Slower? That's a reasonably easy question to address. The strength with which occupants are pinned to the hull (the fake gravity) depends upon the diameter of the ship and the speed at which it rotates. To explore the realism of this, we start with one equation and one definition. The equation for centripetal force is

$$F = \frac{mv^2}{R}$$

where F is the force felt by the object spinning (in our case, the person feeling the artificial gravity), v is the rotational velocity, m is the person's mass, and R is the radius of the spinning section of the ship. We further define Γ as a factor that indicates the number of Gs we wish to emulate. In other words, if we have a ring spinning fast enough to simulate twice the force of Caprican gravity, then $\Gamma = 2$. Obviously, Γ would normally be 1 because we want to simulate one Caprican G. So to express the above equation in terms of G forces, we have

$$\frac{F}{g} = \Gamma = \frac{v^2}{gr}$$

where g is the normal downward acceleration of gravity, 9.81 m/s^2. The miracle of algebra occurs and we have three equations. To find the simulated gravity for a given spin radius and spin rate,

$$\Gamma = \frac{R}{g}\left(\frac{\pi f}{30}\right)^2$$

where f is the rotation rate in RPM and π is the constant 3.1416. To find the spin radius for a desired force of gravity and RPM, we rearrange and use

$$R = \Gamma g \left(\frac{30}{\pi f} \right)^2 .$$

Rearranging again, for a given rotation radius and simulated gravity, the rotation rate in RPM is given by

$$f = \left(\frac{30}{\pi} \right) \sqrt{\frac{\Gamma g}{R}} .$$

So if the radius of *Zephyr*'s ring section is 500 meters, which is close to what it appears onscreen given its perspective to other ships, in order to simulate 1G of gravity the ring would have to spin at 1.33 RPM, or 1⅓ times the rate that a second hand moves around a clock dial. This happens to be very close to what is observed onscreen. What do you want to bet that somebody in the *Battlestar Galactica* visual effects department did this calculation as well?

If all objects have gravity, doesn't *Galactica*, as massive as it is, exert its own gravitational force? Yes. Can that be what keeps everyone glued to the deck? No. *Galactica*'s gravitational field is large, and *Galactica* surely warps the space around it, but nowhere to the extent that would provide artificial gravity similar to that of a terrestrial planet. It wouldn't even provide gravity equal to that of a small asteroid, so the only other reliable way to make artificial gravity is to use science fiction and just make "artificial gravity." This has been the other way to present space travel on a low budget, but it has just as many (if not more) problems as centrifugal ersatz gravity.

How does *Galactica*'s artificial gravity work? We don't know and probably never will, but we can make some calculated guesses as to the requirements and constraints upon its implementation. In chapter 12, "General Relativity and Real Gravity (or the Lack Thereof)," we established that Isaac Newton's Law of Universal Gravitation, expressing the force of gravity between two objects separated by a distance *r*, is

$$F = G\frac{m_1 m_2}{r^2}.$$

(We recognize that General Relativity is a better way to understand the cosmos, but Newton's equation is still used today as a good approximation of gravity in many theoretical applications.) If we take the time to understand why the force of gravity gets weaker as a function of $1/r^2$, then we can understand what might be involved in a practical implementation of artificial gravity.

Let us imagine that any object with mass will emit gravitons: massless particles that travel at the speed of light. Further, imagine that gravitons are the agents that warp space. The more mass in a volume of space, the more gravitons are emitted; the more gravitons in a given volume of space, the greater the warping of space; the greater the warping of space, the greater the gravitational pull. We're not alone in our imaginings; this is how many physicists understand how gravity operates (at a very cursory level, of course). If the model shown to the left is true, then a massive object like a planet or a star would emit gravitons that propagate radially outward from its center.

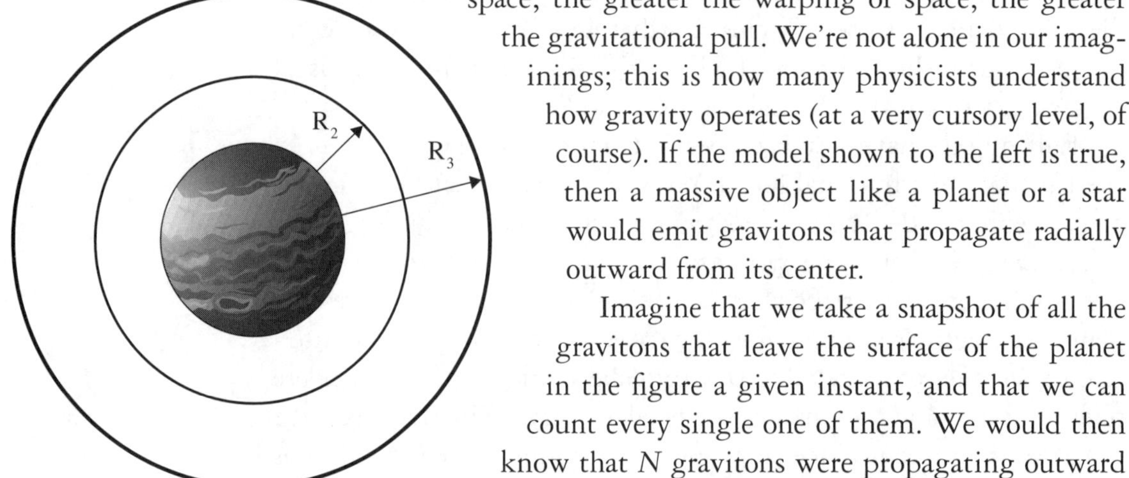

Gravitons emitted from a massive object like a planet.

Imagine that we take a snapshot of all the gravitons that leave the surface of the planet in the figure a given instant, and that we can count every single one of them. We would then know that *N* gravitons were propagating outward at the speed of light in a spherical shell that is one graviton thick. If the degree to which space is warped is determined by the number of gravitons in a given volume, the density of gravitons (σ) for that snapshot in time would be

$$\sigma_{initial} = \frac{N}{4\pi r_{planet}^2}$$

where r_{planet} is the radius of the shell (initially the same as the radius of the planet) and $4\pi r^2$ is the surface area of the expanding spherical shell. When the shell expands to radii R_2 and R_3, the density of gravitons is

$$\sigma_2 = \frac{N}{4\pi R_2^2} \qquad \sigma_3 = \frac{N}{4\pi R_3^2}.$$

If we compare the graviton density when the shell is at R_2 compared to when it first left the planet, we have

$$\frac{\sigma_2}{\sigma_{planet}} = \frac{\dfrac{N}{4\pi R_2^2}}{\dfrac{N}{4\pi r_{planet}^2}} = \frac{r_{planet}^2}{R_2^2}.$$

If $R_2 = 2r_{planet}$, then

$$\frac{\sigma_2}{\sigma_{planet}} = \frac{r_{planet}^2}{\left(2r_{planet}\right)^2} = \frac{1}{4}.$$

If $R_3 = 3r_{planet}$, then

$$\frac{\sigma_3}{\sigma_{planet}} = \frac{r_{planet}^2}{\left(3r_{planet}\right)^2} = \frac{1}{9}.$$

We can now see why the force of gravity decreases as $1/r^2$ and now have an even better understanding and appreciation of Newton's Law of Gravitation! We can now better understand some of the issues behind the creation of artificial gravity.

If the Colonials could somehow generate an artificial gravitational field, they would therefore likely use some sort of device that worked by emitting gravitons. (We have zero idea how to build a controlled graviton emitter, but let us assume that the Colonials know all about them and how to harness them.) The gravitons would have to be emitted in a directed pattern that mimicked planetary gravity. Since gravitons would not carry an electric charge, it would be nearly impossible to create a directed beam like we can do with charged particles (like in a linear accelerator) or photons (as with a laser).

Since gravity's effects diminish as $1/r^2$, an antigrav machine in the bowels of the ship (say near the CIC) would create Caprica-like gravity in the CIC only! Other parts of the ship would experience a gravity gradient—stronger gravity near the CIC, diminished gravity near the bow, stern, and flight pods. Items at the extremes of the ship would (1) feel significantly "lighter" than objects in the center of the ship and (2) would all lean away from the CIC! This is clearly no way to live.

Most likely, the Colonials use tiny graviton generators (we'll call them NAGGs, short for Nano Artificial Gravity Generators) embedded in each deck in each room and corridor. The NAGGs must be able to shape the gravitational field upward so that the gravitational field fills the room without making extremely tall people on the deck below feel as if they were being pulled toward the ceiling. Ideally, the field must be strong enough to keep people glued to the deck, and that means emitting the gravitons in a hemispherical pattern at best. If a NAGG were placed in the center of the room, we would have a similar problem as placing one near CIC, only in miniature. You'd be leaning away from the center of the room, or feel alternating gravity highs and lows while walking down the corridor. If the NAGGs are spaced linearly down the centerline of each hallway, and in a stripe within each room, then the graviton field would take the shape of a half-cylinder emanating from the centerline of the deck.

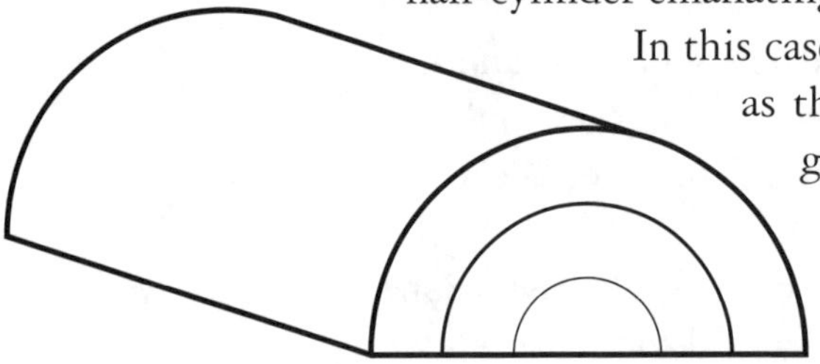

In this case, gravity would trail off as a function of $1/r$ as the further you were from the NAGGs. The good news is that this is an improvement over $1/r^2$. The bad news is that this stifles crew social interaction: everybody walking down a corridor would be leaning ever so slightly away from the hallway centerline.

Gravitons from an artificial gravity generator.

If you put multiple NAGG strips down the corridor, we start to approach something that looks more like a constant field, without the highs/lows of our previous efforts.

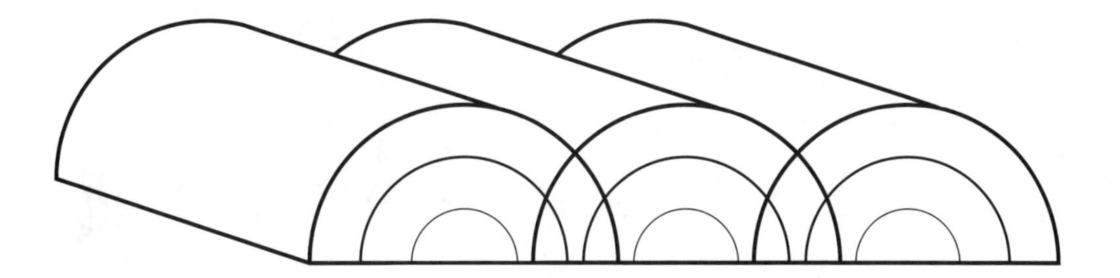

Gravitons from an artificial gravity generator.

Clearly the more NAGGs embedded in the deck, the more planet-like the gravitational field. If graviton generators existed, they would be spread uniformly across any area that required gravity.

All of this assumes that individual graviton generators could create hemispherical patterns. More than likely, the nature of gravitons would radiate in a spherical pattern. This means that if a society were capable of generating gravitons, and were able to make the generators small enough to embed within a deck, the gravity generators in the floor above would pull you upward nearly as powerfully as the ones beneath your feet would pull you down. So instead of the expected situation where the floor for one level is the ceiling for the deck beneath, it's likely that each floor that had NAGGs would be a "floor" in two directions, similar in structure to the hangar bays on the *Pegasus*.

Although we have never had any visual clue that Vipers have artificial gravity, we have clearly seen that Raptors do. If the artificial gravity were engaged, and unless the gravity field were hemispherical, how would a Raptor ever get off the deck? Clearly, for tactical spacecraft like the Raptor, an artificial gravity system would be more of a complication.

The bottom line is that even if artificial gravity could be created, by generating gravitons instead of rotating the ship or undergoing a constant acceleration, the problems are just beginning. Even after a society understands the nature of gravity, something that is currently beyond the ken of Earth physics, there will still be technological hurdles to overcome. In the foreseeable future, and perhaps forever, artificial gravity will belong to the realm of science fiction.

CHAPTER 24

Navigation

Food, water, and fuel shortages. Few, if any, basic services. The constant fear of imminent Cylon attack. Although the Colonials have more than their share of problems, the problem of being "lost" in a Galaxy with over 300 billion stars scattered throughout its 32 trillion cubic light-years might be just as unsettling and as perilous as any of the others. In the episode "Lay Down Your Burdens, Part I," Starbuck tells a Ready Room full of pilots and ECOs that the Twelve Colonies, in particular Caprica, are "nineteen plotted jumps away." This is after a year on the run, one that began with 240+ (presumably somewhat random) jumps in "33." The implication is that although humans have probably explored only a tiny fraction of the Galaxy, there is an infrastructure or methodology in place that allows them to recompute the positions of the Colonies, and that allows them to know their relative position within the Galaxy at any time. How would they accomplish this? How might the Rag Tag Fleet navigate from the radioactive remnants of the Twelve Colonies to Kobol to Dead Earth, and subsequently on to Earth II? The navigation of any vessel, from an automobile to a battlestar, is a process by which a navigator answers three basic questions: "Where are we?," "Where are we going?," and "In which direction do we go to get there?" While these questions sound fairly straightforward initially, interstellar navigation is another very good example of the adage "Anything studied in sufficient detail becomes infinitely complex."

Defining a Coordinate System or Reference Frame

The first step in navigation comes in answering "Where am I?," or determining both your position and velocity—known in interplanetary navigation as your state or state vector—at a given moment in time. When a navigator makes that confident pronouncement, "We are *here at position* X," there is an implied allusion to an even more fundamental issue, "relative to *position* O." A position is meaningful only when given in a specific coordinate system or reference frame (the two terms will be used synonymously).

The definition of a coordinate system starts with, well, duh, a starting point—also known as the origin. This is the central point from which all positions will be measured. For Earth-based coordinate systems, the origin is typically the center of the planet. In the well-known rectangular, or Cartesian, coordinate system, three perpendicular lines, normally called the X, Y, and Z axes, radiate from the origin (three because our universe has three spatial dimensions). Any object's position in space can now be specified exactly by three numbers, called coordinates: the object's distance from the origin as measured on the X axis, the Y axis, and the Z axis. At least, it can be once we take care of one more tiny detail: we must specify the orientation of the coordinate axes.

On Earth the Z axis is synonymous with the spin axis, meaning that it passes through both the north and south poles (so the X and Y axes would both be in the plane of the equator). A line drawn along

Kara "Starbuck" Thrace, Sharon "Boomer" Valerii, and Lee "Apollo" Adama.

Chief Galen Tyrol.

the surface of Earth from pole to pole is called a meridian. By international agreement, the meridian that passes the Royal Observatory in the town of Greenwich, England, is called the Prime Meridian. Now draw a line from the center of Earth to the point where the Prime Meridian intersects the equator, and we have finally defined the location of the X axis for Earth. The good news is that we have now completely specified a three dimensional coordinate system for the surface of Earth and we understand the basics of how one might construct a three dimensional coordinate system.

Unfortunately, in our everyday lives we do not often need three dimensions to locate a spot on Earth's surface; we naturally assume that any place we want to go is actually on the surface. So for most of our Earthly navigation needs, we treat Earth as a two dimensional map. This way, any object's position can then be specified by using just two angles: longitude, a value that can range from 0 degrees to 360 degrees around Earth's equator, and latitude, a measure of the angular distance above or below the equator, ranging from −90 degrees at the South Pole, to 0 at the Equator, to +90 degrees at the North Pole.

Let's determine the location of Universal Studios in Los Angeles, California, where *Battlestar Galactica* was written. If we measure along the equator and determine the angle from the Prime Meridian to the meridian passing through Universal Studios, we determine that its longitude is 118.35 degrees west. If we measure along the meridian passing through Universal Studios and determine its angle above the equator, 34.14 degrees, we have its latitude.

That works for objects on and near Earth, but not elsewhere. Since Earth is rotating, any Earth-fixed coordinate system is rotating as well. This would not be an ideal frame for navigating among the stars, since the coordinates of any star in the sky would vary dramatically over the span of 24 hours. A rotating reference frame like this is called an accelerated or non-inertial frame. For interstellar travel, we want to establish a coordinate system that has axes that are permanently fixed, or at least close to fixed, for very long time periods.

Astronomers use a similar reference frame, which also has two angles akin to latitude and longitude, to locate objects in the evening sky. Imagine that Earth is surrounded by a transparent sphere thousands of light-years in radius. Imagine also that the stars we see in the night sky are affixed to this Celestial Sphere. (Real stars in the real sky move through the Galaxy, but their movement is noticeable only over many human lifetimes; for navigational purposes, the stars are as motionless as the phosphorescent stars children affix to their ceilings.)

The coordinate system used by astronomers starts by projecting Earth's equator onto the Celestial Sphere. This is called the Celestial Equator. If we project Earth's north and south poles onto the Celestial Sphere, we have the north and south Celestial Poles. Like latitude and longitude, positions in the sky are measured with two angles: right ascension and declination. Right ascension is measured along the celestial equator and ranges from 0 to 360 degrees (though astronomers often use 15-degree increments called hours), similar to longitude on Earth. Declination is measured as an angle "north" or "south" of the celestial equator, similar to latitude, and similarly ranges in value from +90 degrees to −90 degrees. What is the equivalent of the Prime Meridian, the arbitrary place where we start counting degrees? By scientific convention, the intersection of Earth's orbit with the Celestial Equator is a point in space near the constellation Aries (called the Vernal Equinox), and is the celestial equivalent of Greenwich, England. Such a two-angle system like latitude and longitude, or right ascension and declination, is also implied throughout the series *Battlestar Galactica*. How many times have we heard Lieutenant Gaeta say, "We have a Cylon Raider, CBDR, bearing 123 carom 45"?*

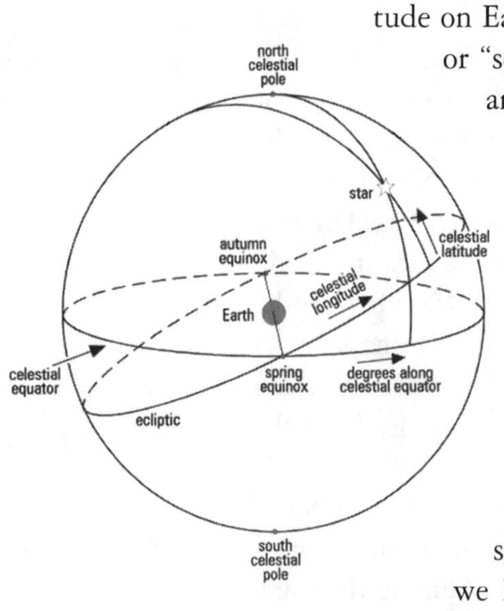

The Celestial Sphere.

Obviously the Colonials do not use degrees as their angular measurement, since we have heard bearings with numbers far greater than 360 on numerous occasions. This simply implies that the angular size of one of their graduations is a fraction of a degree, making them a bit

*Constant Bearing/Decreasing Range, a collision course. The term was first used in the episode "Final Cut."

more accurate than our system. As for distances in space, the Colonials have used three throughout the series. For very close distances in space near *Galactica*, and on the ground of a planet, they have used kilometers. Within a stellar system, the term "SU" was first used in the episode "Captain's Hand," and is presumably of roughly the same scale as an Astronomical Unit (or AU) within the solar system. One AU is colloquially defined as the average distance from Earth to the Sun, about 149 million kilometers, or 93 million miles. Finally, for interstellar distances the Colonials use the same term as terrestrial astronomers: the light-year. Since light is the fastest thing in the universe, it is reasonable that any spacefaring race might use it, in some way, to define vast distances. Since we think that intelligent life as we know it can exist around stars of only a very narrow range of stellar masses (say F5 to K9), and within fairly narrow zones around those, it's reasonable to assume that a "year" defined by the Colonials ranges from 10 months to 15 months. So one Colonial light-year would range from 81 percent to 125 percent of what we colloquially call a light-year.

This coordinate system works perfectly well within the entire solar system because the size of the solar system is tiny compared to the distance of nearby stars. Presumably something like this may work even within the Twelve Colonies. What if we are traveling vast interstellar distances, however? In that case our coordinate system references will appear to shift. We have to look for something else as a basis for our coordinate system.

What about the very center of the Galaxy? We saw in chapter 15, "Our Galaxy," that the solar system orbits the center of the Galaxy every 225 million years, give or take a few million years. Could we use the center of the Milky Way Galaxy as a reference? This is, in fact, one of many coordinate systems used by astronomers today, and is called the galactic coordinate system. If we start with our solar system as the center reference point, the Colonials would obviously start at some point within the Twelve Colonies—then one axis might start at the origin and pass through the center of the Galaxy. Another axis would be perpendicular to the plane of the Galaxy, and the third axis perpendicular to both of them. A slight improvement upon that might be if the initial axis passed either along, or perpendicular to, the long axis of the galactic bar. The position of any object is given as a distance, an angle equivalent to longitude or right ascension that ranges from 0 to 360 in the galactic plane, and an angle like latitude or declination ranging from +90 to −90 degrees above or below the galactic plane. With FTL technology, it's reasonable to think that the colonies may have launched FTL-capable robotic probes to map much of the Galaxy, or even on trajectories perpendicular to the galactic plane, to map our Galaxy from above/below. This may, therefore, sound like a good choice for coordinate systems, and perhaps it would be within the colonies and for journeys to comparatively nearby star systems. But again problems arise. In the plane of our Galaxy there are not only countless billions of stars to obscure the line of sight, there are also vast dark clouds of interstellar dust that inhibit visibility over all but comparatively short galactic distances—a ship that loses track of its references may not end up "lost" per se, but may not know how to get where the crew wants it to be. Perhaps, then, we need to look outside of our Galaxy for a fixed reference frame.

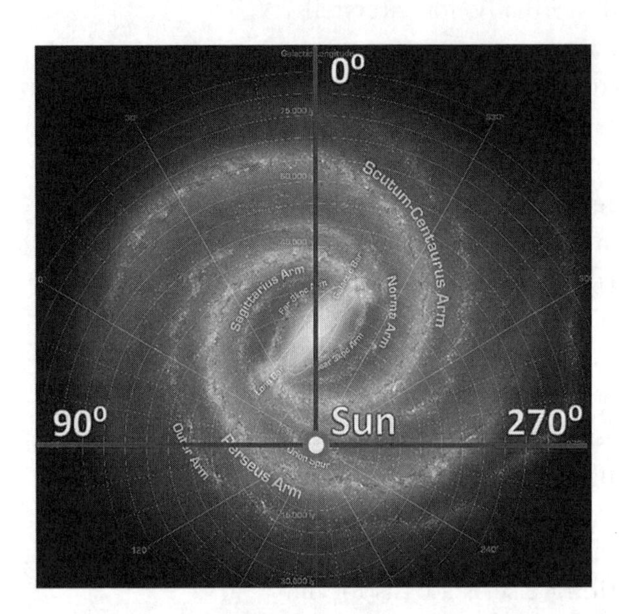

The Galactic Coordinate System.

One fundamental outcome of Albert Einstein's Theory of Special Relativity is that there is no universal standard of rest. As seen from any Galaxy, the universe is expanding uniformly in every direction—so there is no "center of the universe." An implication of this is that we are perfectly justified in saying that our Galaxy, the Milky Way, is fixed and at the center of the universe, and all other galaxies are expanding away from ours. In looking at the universe that way, finding something that is as fixed as stars on a bedroom ceiling is impossible, but finding something that appears to be fixed is trivial. With a few local exceptions (like M31), other galaxies are so very distant that their orientations relative to one another seen from within the Milky Way will never change—not in a hundred thousand human lifetimes, and not from any vantage point. Since galaxies are not concentrated in any area of the sky, we could choose a select number of reference galaxies that are bright enough to be picked out among the stars, and out of the plane of the Milky Way, so they are less likely to be obscured by our own gas and dust. Many spacecraft today have star trackers—a system consisting of a camera that images the field of stars, and a computer that determines which stars within the field of view are those whose positions are stored in an onboard database—to determine spacecraft orientation or attitude. Similarly, it might be reasonable that after every FTL jump, there is an onboard system (or a crew member) that determines the visibility and orientations of select reference galaxies (and perhaps even celestial landmarks), so *Galactica* would always know its orientation with respect to its coordinate system.

In fact, throughout the show we see *Galactica*'s astrogators using simple astrophotographs when calculating jumps. If the astrophotographs are being used also to triangulate the position of *Galactica*, what might they be using as landmarks? Just having a fixed reference frame is a necessary first step, but it is insufficient to answer the question "Where are we?"

Where Are We?

As we looked for very distant objects that are fixed permanently as a basis for a coordinate system, we want to look for comparatively nearby reference points to triangulate our actual position—just as islands, lighthouses, and other landmarks were used by ancient mariners. By determining the relative bearing of three landmarks, navigators at sea "triangulated" their positions. Today's technological variation on the notion of landmarks is the "constellation" of thirty-two satellites orbiting Earth in various orbits, all about 20,200 kilometers above its surface. These satellites constantly broadcast messages with details on their orbital positions and message transmission time. The GPS receiver, which is essentially a small single-purpose radio receiver and computer, records the time the message was received, and uses that to determine the distance to the satellite. It then figures out where the satellite is in its orbit, and calculates the many places on Earth where you would be able to see that particular satellite at that particular distance at that particular time. By using four or more satellites, those many places can be narrowed to a single place, usually to within 30 feet.

In the early 1970s two scientists found a sort of naturally occurring GPS system that triangulates Earth's position in space. NASA was about to launch *Pioneer 10*, the first space probe to the planet Jupiter, and the first man-made object specifically designed to leave the solar system. Since *Pioneer 10* was Earth's first emissary to the stars, the Cornell planetary scientist Carl Sagan petitioned NASA to include a letter of introduction, from Earth to the cosmos, on the spacecraft—one that included directions to the spacecraft's planet of origin. In only a few weeks, Sagan and Dr. Frank Drake, then a professor of astronomy at Cornell, designed a 6-x-9-inch plaque of gold anodized aluminum that carried an engraved calling card from Earth.

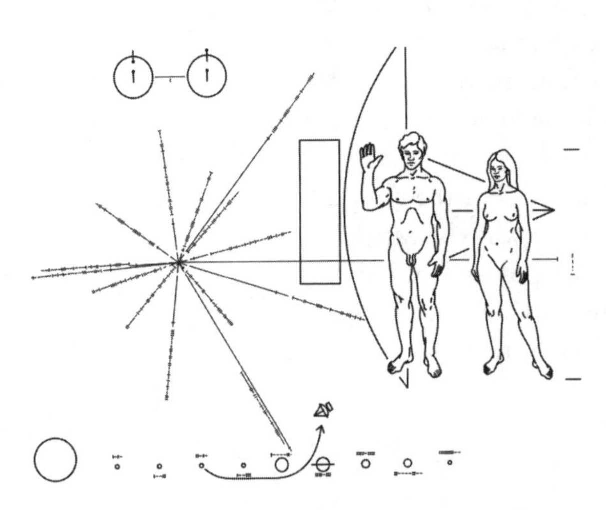

The *Pioneer* Plaque.

The plaque was unveiled at a NASA press conference shortly before the launch, and interest and uproar were instantaneous. "Nudes and Map Tell about Earth to Other Worlds," ran the headline in the *New York Daily News*. Various newspapers across the country ran articles about the plaque, but displayed family-friendly versions of the plaque, selectively edited for public consumption. Yet the newspapers were not upset with the multiline "burst" in the left center of the plaque—even though for the people of planet Earth it had the potential to be the most dangerous piece of information in the entire message. That collection of lines is a map telling space aliens how to find Earth. In fact, some scientists who argued against the plaque's inclusion on the *Pioneer* spacecraft in the first place claimed that it sent out a very clear message: "Here's what's on the menu, and here's a map to the restaurant."

How does the plaque reveal the position of Earth? Instead of orbiting satellites like GPS, Sagan and Drake used something more exotic and naturally occurring. Look very carefully at their map: the center of the map represents the location of Earth, as seen from above the North Galactic Pole. Each of the radial lines represents the angle and distance to fourteen different pulsars. Further, each of those lines ends with approximately 30 binary digits. The 30 binary digits are durations measured as units of time (in 1,420 million cyles per second or, more correctly, its inverse, 7 ten millionths [7/10,000,000] of a second per cycle), and they describe periods of several seconds to several milliseconds. In other words, these lines convey information on the rotation rates and rate of slowing of known pulsars at the time of *Pioneer*'s launch.

Sagan and Drake decided to use the rotational signatures of fourteen different pulsars, hoping that any civilization that could retrieve a small probe from deep space could also easily determine which fourteen pulsars had those particular rotation rates. By working backwards, any civilization that finds the plaque would be able to say that those fourteen pulsars had those specified rotation rates at such-and-such a period in time, and were visible at the given angles only at such-and-such a position in space. Sagan claims that if the extraterrestrials have a good pulsar database, the map will pinpoint 1971, the specific year the plaque was designed. (Without a good database, it'll only specify the twentieth century.) Likewise, the position of Earth could be approximated to within about 60 light-years, and our exact planetary system pinpointed by the drawing of the solar system across the bottom of the plaque.

Could the Rag Tag Fleet navigate across the Galaxy using pulsars as landmarks? Perhaps while the Fleet was close to the Twelve Colonies, but certainly not much farther. To triangulate her position, *Galactica* navigators would need to have landmarks visible over fairly vast distances. When Sagan and Drake used pulsars as landmarks in 1971, pulsars were fairly recent discoveries. We have since learned that pulsars are only seen to "pulse" from vantage points situated on a cone centered on the pulsars' rotational poles. Move very far away from Earth, and perhaps none of the pulsars on the *Pioneer* plaque may appear to "pulse." Sagan and Drake had a good idea, but it probably would not help our brave Colonial heroes navigate across the Galaxy.

What about pattern-matching visible stars and constellations? We can rule this out for all but the shortest of journeys. Our current spacecraft determine their orientations using star imagery, but they only operate within the solar system, where the stars look the same from Neptune as they do from Earth. We tend to forget that the stars of our well-known constellations are three-dimensional; many of the stars are at vastly different distances. Over journeys of many light-years, constellations change their apparent shapes dramatically, rendering them all but unrecognizable the further you get from home. But if we know what the star patterns look like from our destination, more and more constellations will become visible as we get closer to our target. Perhaps, as Lieutenant Gaeta did in "Revelations, Part II," we might arrive at our destination and announce, "Visible constellations are a match!"

Can we just match observed stars to stars in a large database? It sounds possible, but again, the answer is no. Having a computer program match visible stars to a database of known stars is a type of problem that computer scientists call an NP-complete problem (where NP stands for "nondeterministic polynomial"). NP-complete problems have the nasty property that, even for small input sizes, they take the fastest computers a sizeable fraction of the age of the universe to solve. Recall that Aaron Doral said that *Galactica*'s computers "barely deserve the name," and you have an unrealistic choice of navigational aids. Further, almost any star projected against the plane of the Galaxy will be difficult to detect because of the sheer number of stars visible.

Some very bright and very rare or very unusual stars may make good landmarks, but for only comparatively short segments of our journey. Red supergiant stars, for example, are rare but very bright, so they could be seen for reasonably long distances. There are also variable stars—stars whose brightness oscillates from bright to dim back to bright over the span of days or weeks—that would make good short-term landmarks. The period over which a variable star's brightness changes is unique for each star, making them interesting targets as landmarks. It is unlikely, though, that the Colonials have a comprehensive database of every red supergiant or variable star in the Galaxy. Even if they did, views to these stars would eventually become obscured by the vast amounts of dust present in the plane of the Galaxy and swamped amidst the countless stars that compose the disk of the Milky Way. These would make excellent landmarks, but for segments of the journey only.

What about the center of the Galaxy, the galactic bulge? To triangulate the Rag Tag Fleet's position within the Galaxy, this would be an excellent starting point. One of the ways in which modern spacecraft determine their position is by a process called optical navigation, or OPNAV. Let us use the *Cassini* spacecraft orbiting Saturn as an example. By taking an image of one of Saturn's moons, and comparing that to the fixed backdrop of stars behind it, navigators can narrow down the *Cassini*'s position. As *Galactica* moves through the Galaxy, the galactic core would appear to move against the backdrop of the distant galaxies. Although it is in the plane of the Galaxy, along with countless billions of stars, the core can still be resolved. Even though there are also vast clouds of dust in the plane of the Galaxy that obscure line-of-sight tracking of objects like individual stars and nebulae over large distances, electromagnetic radiation in the infrared portion of the spectrum passes through dust more readily than does light in the visible portion of the spectrum. Therefore, if *Galactica* has an infrared telescope, they could resolve the galactic core from any place within the Galaxy. By determining the position of the core relative to the galaxies of our reference frame, like a modern spacecraft OPNAV, *Galactica* could determine their relative bearing to the core of the Galaxy. This would not pinpoint *Galactica*'s position within the Galaxy, but it would be an excellent first step.

Are there any other landmarks, like the galactic core, that appear to move relative to our fixed backdrop of distant galaxies? As a matter of fact, there are. Recall from chapter 15, "Our Galaxy," that the Milky Way is surrounded by a halo of globular clusters. These structures are not only large enough to be seen over very great distances, but most are out of the plane of the Galaxy, and hence are easier to see from a spacecraft immersed within the Galaxy. They are also close enough to the Galaxy that as *Galactica* moves through the Galaxy, they would appear to move relative to our inertial reference frame. As galaxies go, the Milky Way is a fairly large one. Orbiting the Milky Way are other, smaller galaxies. In particular, there are two irregularly shaped galaxies that can be seen only from the southern hemisphere. They appear as two wispy clouds in the night sky, and are called the Large and Small Magellanic Clouds. These, along with other "nearby" galaxies, could serve the same function as the globular clusters that ring the Milky Way—they would make excellent navigation aids, they are visible over vast expanses of our Galaxy, and they would move against the fixed backdrop of distant galaxies as a spacecraft moves through the Galaxy.

The universe has, in fact, provided *Galactica* and her Fleet with a relatively easy way to navigate across the light-years from the Twelve Colonies to Earth. Perhaps when we see negative star images on *Galactica*'s plotting table, we're seeing the Colonial equivalent of an OPNAV—images taken to locate distant reference galaxies or closer landmarks. With distant galaxies chosen to form a fixed, or inertial, reference frame, and by using the galactic core, globular clusters, nearby galaxies, and interesting stars like red supergiants and variables as landmarks, getting lost in the vast expanse of the Milky Way turns out to be toward the bottom of the Rag Tag Fleet's list of concerns.

But how can we forget the episode "Home, Part II"? An away team finds the Tomb of Athena on Kobol and activates a holographic planetarium show, displaying the night sky as it appears from Dead Earth. Lee Adama points to a fuzzy patch of light, identifies it as the Lagoon Nebula, and claims, "At least now we have a map. And a direction."

Like constellations, nebulae have dramatically different appearances when seen from different vantage points. This is a little helpful as a navigation tool, but only a little. The Lagoon Nebula would have the same appearance as it does from the solar system for a fairly narrow range of viewing angles. That's the good news. On the other hand, recall that although visible from Earth, the Lagoon Nebula is really not in the solar system's neighborhood, it is 4,100 light-years—820 red line jumps—away. Let's assume that the Lagoon Nebula appears, more or less, the same within a cone of 5 degrees from a line adjoining the solar system and the nebula, which is a conservative estimate.

The volume of a 5-degree cone 4,100 light-years in length is 552 million cubic light-years. Given that there are approximately 300 billion stars in the Milky Way, that means there would be approximately 5.2 million stars in that cone. If 10 percent of those are stars that could have habitable planets, that leaves only 520,000 potential locations for Earth. Although that is a much smaller search space than 300 billion, it is still hardly a manageable number. The Colonials would need a lot more information to find Earth, which they obviously get ultimately.

A more intriguing question is "How did Apollo recognize the Lagoon Nebula in the first place?" Perhaps from sacred scripture? Certainly if it's a known object in the night sky of the Twelve Colonies, and it has roughly the same appearance there as from Dead Earth, then the Twelve Colonies, Dead Earth, and the Lagoon Nebula are all, more or less, along a straight line in the Galaxy. In fact, since the Lagoon Nebula and the galactic core are in the same portion of the sky as seen from the solar system, then the galactic core, the Lagoon Nebula, the solar system, and the Twelve Colonies are all close to being collinear. Similar to the previous argument, this narrows the search space dramatically, but still leaves a large number of stars through which they would need to comb. If the Twelve Colonies were even twice the distance from the nebula as Earth, then the cone in which they would need to search for Earth has a volume of 4.4 billion cubic light-years, and it would contain approximately 41 million star systems. If 10 percent of those were habitable, the Fleet would be looking at a search space of over 4 million stars. "Needle in a haystack" is trivial in comparison.

CHAPTER 25

Battlestars, Vipers, and Raptors

A modern carrier battle group consists of an aircraft carrier and escort ships—frigates, destroyers, and cruisers—that act as both anti-air and anti-submarine shields for the carrier and supply ships of the battle group. Although the Colonial warships seen throughout the series have been battlestars (including *Galactica, Pegasus, Columbia, Valkyrie*, and *Yashuman*), the Twelve Colonies obviously operated smaller combatant ships like those in today's battle groups. In "Razor," although the battlestars *Bellerophon* and *Ramses* were docked at the Scorpia shipyards with *Pegasus*, several smaller vessels were also present. Further, in the episode "Daybreak," William Adama says, "I've commanded two battlestars, three escorts before that."

Since the initial design of a military vessel represents a large fraction of the cost of a ship, it makes economic sense to build multiple copies of the same design. Therefore the navy does not simply build the destroyer U.S.S. *Arleigh Burke*, it builds sixty-two *Arleigh*

Lee plays a game of triad.

Lee pins Starbuck's picture on the memorial wall.

We'll be in too close for nukes. Same goes for missiles. This'll be strictly a gun battle. Like two old ships of the line slugging it out at point blank range. Let the gun captains know their job is to fire and keep firing until they run out of ammo. Then they should start throwing rocks.

—Admiral Adama, *Battlestar Galactica*, "Daybreak, Part II"

Burke–class destroyers all having (roughly) the same design, The same is apparently true of the Colonial Fleet. In the episode "Pegasus" we learn that *Galactica* is a *Jupiter*-class battlestar, while *Pegasus* is a much newer *Mercury*-class.

It's tempting to think of a battlestar as simply a space-based aircraft carrier. However, since battlestars are heavily armored and are also armed with serious firepower, they are more akin to a hybrid aircraft carrier/battleship.

To defend herself, *Galactica* has multiple batteries of kinetic energy weapons, or KEWs. Remember that at the beginning of the book, we described kinetic energy as the energy a body possesses because of its movement? In that way, slingshots, catapults, guns, rifles, and cannon are all kinetic energy weapons; they cause damage simply because of the combined mass and speed of their projectiles. We also said that kinetic energy was easy to transfer

The Battlestar *Pegasus*.

from one body to another. When these projectiles transfer their huge kinetic energy to an enemy target many meters or even kilometers away, some of the kinetic energy becomes heat, and some of the kinetic energy is transferred directly into the target, either blowing a hole in it or imparting momentum to it (or both).

As we previously discussed in chapter 9, "Energy Matters," the most commonly used unit of energy is the joule. A two-kilogram mass traveling at 1 meter per second has a kinetic energy of 1 joule. A 100-kilogram (about 220 pounds) NFL halfback running at 6 meters per second has 1,800 joules of kinetic energy. By comparison, a round of .223 caliber ammunition, used by the U.S. military's standard issue M16 rifle, has over 1,750 joules of kinetic energy. A single bullet from an M16 hits as hard as being flattened by an NFL halfback. That's the whole point behind a bullet. When the bullet strikes the target, its small size helps to focus the kinetic energy at the point of impact, giving this tiny hunk of lead a punch like a laser beam.

Battlestar rail guns.

While bullets, cannonballs, and other forms of nonexplosive projectiles can be considered KEWs in the broadest sense of the term, in modern terminology the term KEW has a narrower connotation. Kinetic energy weapons are designed to launch projectiles that impact a target at incredibly high rates of speed. Battlestars are equipped with rail guns that fall into this category. A rail gun uses magnetic fields to accelerate a metallic projectile. In 2008 the U.S. Navy successfully tested a rail gun that accelerated an eight-pound projectile up to seven times the speed of sound. The navy currently plans to use rail guns that fire projectiles at eight times this energy when they finally become operational within the Fleet—the equivalent energy of launching our NFL running back at the muzzle velocity of our M16. Clearly, KE weapons don't have to be extremely complex to be extremely effective; rail guns deployed in the real world within the next few years would be powerful enough to pierce the armor of even *Galactica* herself.

Galactica is also armed with nuclear weapons, but as counterintuitive as it might seem, a metal rod accelerated to high velocity is likely to be a far more effective weapon against an armored spacecraft than a nuclear warhead. As we saw, *Galactica*'s armor could stop all, or nearly all, of the explosive force of a nuke from doing damage to the crew or to the ship itself.

The ionizing radiation released by a nuclear weapon, though, generates a burst of electromagnetic energy that can burn out unshielded electronic circuits. A battlestar's electronics would partially be protected by its metallic hull, but sensors on the outside of the ship would be particularly vulnerable. A Cylon ship, in which a large percentage of the ship's structure is organic, would likely suffer more damage from thermonuclear radiation than a battlestar.

Finally, nearly half the energy of a nuclear weapon detonation is the blast wave that propagates away from the explosion at nearly the speed of sound. This is actually a good thing, because in space no one can hear your nuke going off. Without any air, there is no medium to propagate a blast wave. Nuclear weapons would lose much of their effectiveness in space because of this.

Although a battlestar has a formidable array of weapons, it is her fighters that are her main offensive punch. On aircraft carriers, airplanes are launched from the flight deck using steam-driven catapults. The aircraft crew locks the aircraft's nose wheel onto a piston in the deck. An enormous amount of potential energy—in the form of steam pressure—builds up behind the piston. When the pilot signals for takeoff, the piston is freed and the steam pressure becomes kinetic energy driving the piston (and the attached aircraft) down the carrier's launch rail. Given this burst of speed, the aircraft is able to gain enough lift to achieve flight. The system is then rewound, a new aircraft is mounted, and the process begins again with the buildup of more steamy potential energy.

> **A battlestar's whole purpose is to launch Vipers.**
> —Specialist Dealino, "Daybreak, Part I"

One of the drawbacks of this system is that each catapult shot uses up to 614 kg of steam. This may work well on a ship surrounded by water, but on a battlestar in space, where H_2O is a precious commodity, spacecraft would have to launch using another method. Like the next-generation aircraft carriers currently under construction, the catapults in *Galactica*'s launch tubes—magcelerators, or

Vipers launching from a battlestar.

"mag cats"—are magnetically driven. Using a concept similar to that of rail guns, magnetic fields are used to propel a metal shuttle down a launch rail. When a Viper is attached to the shuttle, the fighter is flung into space.

A noticeable difference between the 1978 *Battlestar Galactica* and the reimagined incarnation is the procedure for launching Vipers. A constant from the original series was Corporal Rigel's calm and steady voice: "Core systems transferring control to Viper fighters. Launch when ready."

In the reimagined *Battlestar Galactica*, a launch officer goes through the same formal checklist every time a Viper is launched: "Viper two-eight-nine/*Galactica*, clear forward, nav-con green, interval check, thrust positive and steady. Mag cat engaged. Good-bye, Starbuck." In the original series the decision of when to launch rested with the pilot.

In the reimagined series the decision if and when to launch rested with the launch officer, which is deliberately more like the way a plane is launched from a carrier today. It also made more sense from a physics standpoint; in the original series the energy to launch a Viper came from within the Viper itself, thus consuming valuable fuel. In the reimagined series, *Galactica* brings the Viper up to launch speed externally, a much more energy-efficient way of doing business.

Vipers

If *Galactica* is an aircraft carrier in space, then Vipers are her fighter aircraft—single-seat multimission fighters used primarily for air and space superiority, but with limited ground-attack capability. Like today's fighter craft, they also have limited electronic countermeasures (ECMs) capability as well, as shown by the small pod mounted on the vertical stabilizer.

> **This is the Mark II Viper. It's nimble as a jackrabbit, and anyone not paying attention is likely to become a pile of muck that needs to be hosed out of the cockpit by the chief of the deck.**
>
> —Lt. Kara "Starbuck" Thrace, "Act of Contrition"

A purely space-based fighter would not need to be shaped aerodynamically, since the lack of atmosphere makes aeronautical control surfaces

A Mark VII Viper.

like wings, elevons, a tail, or a rudder meaningless. The Viper, however, is capable of flight within the atmospheres of different types of planets and moons. Since the Viper's engines aren't necessarily "air breathers"—they don't intake air surrounding the craft for use as an oxidizer to generate thrust as a jet engine does—they can fly in just about any atmosphere. Each atmosphere does present its own peculiarities and challenges for the maintenance crew, however.

Although we've seen Vipers deploy air-to-ground munitions ("The Hand of God" and "Exodus, Part II") and air-to-air missiles ("Blood on the Scales"), a Viper's main weapons are its twin wing-mounted cannon. In the episode "Epiphanies," in which Chief Tyrol and his flight crew discover that somebody has been tampering with Viper ammunition, we see that Viper KEWs appear to be belt-fed medium-caliber weapons using chemical propellant to fire solid or explosive rounds. They are very similar to the 20 mm rounds fired by the M61 cannon mounted on all U.S. military fighter aircraft today, with one obvious implied exception. Since Vipers operate in space, the casing for each round must not only contain a propellant like gunpowder, it must also contain an oxidizer. Some ammunition being developed today has both propellant and oxidizer molded to form a temporary case, which is "burned" when the round is fired.

For *Battlestar Galactica*, executive producers Ron Moore and David Eick made the decision that Colonial armaments would fire projectiles instead of the more traditionally science fictional lasers. Those white flashes that some people think are short laser pulses are really tracer rounds—bullets coated with a pyrotechnic material that burns when the round is fired, glowing very brightly. By putting in one tracer for about every ten regular bullets, the shooter can see where the bullets are going and can more accurately adjust fire.

The producers' choice to endow Vipers, Colonial Marines, and battlestars with projectile weaponry was initially met with cries of "Luddites!" from fans when the miniseries initially aired. For many fans, lasers simply seem more . . . lethal than a high-speed chunk of lead or depleted uranium. If you still feel that Vipers could have been made to seem more capable, let's examine how efficiently both types of weapons deliver lethal energy to their targets.

Most common lasers operate in a similar fashion: excite atoms or molecules—electrically, thermally, or chemically—and they radiate photons of a given energy. Recall that "energy" and "color" are synonymous—the "color" of the photons is determined by the substance being excited. Now generate that light within a cavity that has a 100 percent silvered mirror on one end, and a 99 percent silvered mirror at the other, called a resonator chamber. The light that is reflecting back and forth, into the already excited atoms, generates even more photons of the same energy. This is called stimulated emission. The 1 percent of energy that "leaks" through the 99 percent silvered mirror is your projected laser beam.

In comparing directed energy weapons to KEWs, let's use the 20mm high-explosive round fired from an M61A1 cannon—the type mounted on most U.S. fighter aircraft—for our baseline example of a KEW. The shells fired from an M61A1 have a muzzle velocity of 1,050 meters per second (or about 2,350 miles per hour). At roughly 180 grams, or 0.18 kg, each round has a kinetic energy of nearly 100,000 joules. To transfer that much energy in a one-second burst, a laser would have to be in the 0.1-megawatt, or 100-kilowatt, range. The U.S. Navy is currently considering installing 100-kilowatt lasers aboard the next-generation destroyer, the DDG-1000 or *Zumwalt* class, to destroy inbound anti-ship missiles. A laser in the 100-kilowatt range is considered the minimum power to be considered a "weapons-grade" laser.

Modern fighter aircraft, and anti-ship KEWs, can fire six thousand rounds per minute, or one hundred rounds per second. This is dramatically higher than the rate of fire that Vipers have displayed onscreen. To equal the firepower of an M61 cannon at a hundred rounds per second, a capability that U.S. military fighter aircraft have had since the late 1950s, would require a laser in the 11-megawatt range.

The U.S. Missile Defense Agency has developed and tested a 1-megawatt laser, part of a system designed to destroy inbound ballistic missiles, but these lasers are mounted within a specially designed 747 and are far too bulky to mount aboard a craft the size of a Viper. Additionally, the power levels within the resonators of lasers like these are so high that their mirrors are always in danger of self-destructing. The slightest flaw in the mirror's surface—the slightest blemish or even a dust particle—can cause the mirror to absorb the laser energy rather than reflect it, and melt.

Further, recall that Vipers are supposed to be able to operate both in space and within an atmosphere. That inflicts even more design constraints on our directed energy weapons. If the laser is powerful enough, then one must worry about a phenomena called blooming, where the beam ionizes the air through which it travels and creates an opaque plasma that actually dissipates the beam. The higher the energy, the greater the blooming and the more energy dissipated. For a laser to be effective within an atmosphere, it would have to be "tuned" to the particular wavelength for which the atmosphere is most transparent.

Even assuming the Colonies have an advantage over the U.S. military in the realm of technological sophistication, it would take significant engineering and scientific advances to create a laser that has the destructive power that current air-to-air kinetic weapons have had since before the Vietnam War. Perhaps the decision of the *Battlestar Galactica* producers to portray KEWs on their Vipers and battlestars was not a poor choice after all.

DIRECTED ENERGY WEAPONS

Directed energy weapons, such as lasers, are seldom directed well on TV and movies. Either they're *Star Trek*-like phasers, which shoot long, straight beams of energy that are visible in daylight, or they're *Star Wars*-type blasters, which shoot short tracerlike bolts of energy that are also visible in daylight. *Battlestar Galactica* (the new version) got them right by not using them at all.

When we speak of directed energy weapons, the energy being directed is electromagnetic radiation, which travels at the speed of light. Therefore, when a directed energy weapon like a laser is fired in a close-range battle, like within a room or corridor, the energy beam would appear to "connect" from weapon to target as soon as the shooter pulled the trigger, then instantly disconnect when turned off. This is exactly what you see with a laser pointer: as soon as the button is pressed, the laser light appears on the wall, the viewscreen, the ceiling, or the family cat. What you would *not* see is the traditional "bolt" of energy—appearing much like a solid tracer round—like we saw from all the spacecraft in the original *Battlestar Galactica* (and *Star Trek, Star Wars, Stargate*, and *Farscape*, all of which fall short on this point).

Then again, all of this presupposes that you could see the beam at all! In order for you to see anything, light from that object has to interact with a sensor (that is, your eye). The beam of a directed energy weapon cannot be visible unless some part of the beam makes it into your eye. The only way that is going to happen is if you're shot in the eye, or if there is a medium that will scatter the beam, like dust or smoke particles and/or water droplets in air. Rock bands have laser light shows at concerts because they know full well that the hall will be filled with tobacco (or other plant matter) smoke to scatter the beam and make it visible. If you've attended a laser light show—the kind often held in planetaria or other venues that ostensibly do not allow smoking—the beam is nearly impossible to see, except where it reflects off the ceiling.

In fact, perhaps one of the best examples of a realistic portrayal of directed energy weapons was the hand weapons and rifles used by both the Colonial Warriors and Cylons in the original *Battlestar Galactica*. Captain Apollo draws his weapon, aims, pulls the trigger, and instantaneously we see the exploding plasma cloud on the Cylon Centurion at whom he had aimed his weapon. Cylon weapons were depicted in the same way. So that "cheesy" 1970s show is actually a good example of science fiction tech done right, at least in the way it was shown to act onscreen.

Raptors

They have been used for close air support, personnel shuttles, troop transport, commando insertion, reconnaissance, and search and rescue (SAR). Raptors are formidable attack craft—in "Exodus, Part II" and "Daybreak," they carried impressive amounts of armaments ranging from their internal cannon to bombs, rocket launchers, even nuclear weapons. All those varied missions notwithstanding, the primary role of the Raptor is as a command, control, and communications (C^3), electronic warfare (EW), and electronic countermeasures (ECM) platform. In his online blog, Ron Moore wrote: "I never really thought of the Raptor as a transport, I usually thought of it as analogous to the Navy's EA-6 Prowler (a variant of the A-6 Intruder popularized in 'Flight of the Intruder')."

Once again the Raptors get the hard work.
Lt. Karl "Helo" Agathon, "Daybreak"

A Raptor.

Similar to the navy's Prowler, which has a pilot and three electronic countermeasure officers (ECOs), the Raptor has a pilot and a lone ECO. As an EW and ECM platform, Raptors are equipped with a full suite of integrated electronic countermeasure and threat-monitoring instruments, as well as communications drones, decoys, chaff launchers, and flares.

Boomer: Captain, I have two communications pods left, sir, but that's it. No jiggers, no drones, no markers, nothing.

Lee: Well, at least you've still got your electronics suite . . .

In a moment that the writers Bradley Thompson and David Weddle intentionally meant to invoke images and memories of the catastrophic fire aboard the aircraft carrier USS *Forrestal* on July 29, 1967, it was a Raptor communications drone, not a missile, that shot across the hangar bay and killed thirteen pilots in the episode "Act of Contrition."

Just like the navy's EA-6B Prowler (or its successor, the EA-18G Growler), Raptors normally accompany a flight of fighters on missions, or sorties, to provide EW and ECM support. The many and various aspects of EW are detailed in chapter 26, "Toasters and Jam: The Complexities of Electronic Warfare," and this was, in fact, the first role in which we saw a Raptor acting in the miniseries. We see them in numerous roles later; they are the workhorses of the Rag Tag Fleet.

CHAPTER 26

Toasters and Jam: The Complexities of Electronic Warfare

Electronic warfare (EW) is the ultimate military cat-and-mouse game. It wins battles, sometimes wars, before they begin, and its importance has been largely overlooked—certainly underestimated—in both television and film. This is, perhaps, because "military action involving the use of electromagnetic and directed energy to control the electromagnetic spectrum or to attack the enemy" doesn't initially sound like compelling drama. But it can be. Many TV series that allude to EW at all use it as little more than a MacGuffin (for example, "Drat! Our feared and hated enemy is jamming our distress call.").

The complexities of EW can take entire books to describe thoroughly, but it boils down to one thing: in armed conflict, the combatant who can gain advantage in their ability to utilize the EM spectrum, and who can gather more information about their adversary, has an overwhelming advantage—irrespective of who has the biggest guns.

Samuel Anders.

Starbuck and Athena in "Unfinished Business."

Electronic Detection

DRADIS (Direction, RAnge, And DIStance) is similar to modern-day radar (RAdio Detection And Ranging) in that an antenna sends out a pulse of EM radiation in the radio portion of the spectrum. (From this point on, we'll use DRADIS and radar synonymously.) Many types of matter, particularly the kind of metals used to make ships and planes and spacecraft, reflect that radiation, and a portion of that reflection travels directly back to the source. The reflected pulse is detected by the source antenna, and the round-trip travel time of the pulse is measured. From this, and the position of the signal, the distance and direction to the reflecting object can be calculated. Sonar (SONic detection And Ranging) works in a similar way, but instead of broadcasting electromagnetic energy, sonar works by generating acoustic

Adama and the DRADIS.

pulses, or pings, and "listening" for reflections from submerged—or partially submerged—objects. Natural sonar is used by bats and most cetaceans, and is perhaps the original form of "electronic warfare."

DRADIS, like radar and sonar, is depicted on *Battlestar Galactica* as an active form of electronic warfare in that the ship or missile using its sensor to locate the enemy has to transmit energy into its environment. Radar comes in two basic types: search and fire control. Search radar is used to detect the presence of, and provide bearing, range, and velocity information on, other vessels both friend and foe; fire-control radar is used by a weapon system to acquire and/or home in on a target.

An advantage to a system like DRADIS is that once the system senses a pulse of energy reflected from an enemy ship, its existence has been revealed, as has information on its range and direction. One drawback, however, is that an enemy can "listen" passively for a DRADIS pulse. As a pulse of EM radiation propagates away from its emitter, its power dissipates. When that pulse reflects off an enemy vessel, the process is never 100 percent efficient; only a fraction of the incident pulse is reflected, and it is attenuated even more. Then the reflected pulse dissipates still further during the return trip back to the receiving antenna—which is usually, though not always, synonymous with the emitting antenna. An adversary, therefore, can more easily sense the radiation from your DRADIS than your own system can. For example, a ship on the ocean can see a lighthouse beacon long before the lighthouse beacon can illuminate the ship. Radar and DRADIS work the same way: we can spot a radar signal many miles (or dozens of miles) before the radar signal can spot us. In short, while active DRADIS has its uses, it is also a beacon to the enemy saying, *Here I am, come shoot me!* To exploit this type of vulnerability in battle the U.S. military has had in its operational arsenal since 1985 the HARM (High-speed Anti-Radiation Missile)—which has been superseded by the Advanced Anti-Radiation Guided Missile (AARGM). The HARM was designed to detect an enemy radar emission, then home in on and destroy the emitting antenna. These are typically used in the first wave of an attack to destroy enemy anti-aircraft batteries, but can also target radar jammers or even communications installations. In summary, although Vipers, Raptors, and even the Big G herself are portrayed as always using active DRADIS, in reality the choice to go active would be a complex one and would vary dramatically depending upon the tactical situation.

Some types of missiles have radar emitters onboard and can also use active homing to find their intended target. We saw in the *Battlestar Galactica* miniseries that some of the Cylon missiles use active homing—when missiles locked onto both Sharon's Raptor and Colonial One. There are actually three basic methods that missiles of today employ to home in on their targets: active homing, semi-active homing, and passive homing. Many of today's missiles are "smart" enough to switch between different modes of homing, using the method most appropriate for the moment. That did not seem to be the case with Cylon missiles over the run of *Battlestar Galactica*. Did the Cylons lack the technical know-how? Unlikely. It was more likely the case that the Cylons simply didn't think to do this.

A weapon that employs semi-active homing tracks a reflected radar signal broadcast by a separate emitter—often that of the air/spacecraft that fired it. There was an excellent example of this in the second-season episode "Fragged." The Cylon DRADIS on Kobol that Chief Tyrol wanted to destroy was obviously used to both acquire and guide missiles to a target. The landing party destroyed the Cylon DRADIS just as the Cylons fired missiles at *Galactica*'s search-and-rescue Raptors flying overhead. As soon as the DRADIS was destroyed, the missiles lost their lock on the Raptors, allowing us to infer that the Cylon missiles employed semi-active homing. The existence of missiles like the HARM can make the decision to fire weapons with semi-active homing a dangerous prospect—since a HARM missile homes in on the source of the radar—because in the case of semi-active radar, that source may be the targeting aircraft. This simply underscores the risks that exist on the battlefield when switching on any kind of emitter.

In the case of passive homing, a missile tracks the EM radiation emitted by the enemy. Any hot object emits radiation, even human beings. Heat-seeking missiles home in on the infrared radiation from the adversary's hot tailpipe and are usually fired while in a tail chase. Human beings, while not as hot as a spacecraft tailpipe, also emit infrared radiation. This is the portion of the electromagnetic spectrum to which night-vision scopes are sensitive.

Another form of active electronic warfare used in *Battlestar Galactica* is IFF, an abbreviation for "Identification Friend or Foe." IFF is a system first developed in World War I, used by military to identify friendly forces and determine their bearing and range. Real-world IFF systems use an encrypted "question-and-answer" protocol between two craft. If the "questioned" craft responds correctly, it is identified as friendly. A craft not replying correctly can be classified as suspicious, but not instantly be designated as a foe, since there may be many reasons for an incorrect response—like battle damage, for instance. IFF hardware can also be integrated into fire-control and missile systems to reduce the likelihood of friendly fire accidents. In "Kobol's Last Gleaming, Part I" Lieutenant Gaeta was able to determine that the Cylon transponders found within the Fleet, including one attached to the DRADIS display, emitted an IFF pulse when they got within range of one another. This fact was later used to fool the Cylon armada at Kobol into believing that Boomer's Raptor was "friendly."

Electronic Countermeasures

It may have started with the Trojan horse, or even earlier, but deception and warfare have always gone hand-in-hand. The U.S. Department of Defense (DoD) defines deception as "those measures designed to mislead the enemy by manipulation, distortion, or falsification of evidence to induce the enemy to react in a manner prejudicial to the enemy's interests." Just like the Trojan warriors evaded detection over three thousand years ago, modern-day radar and other detection systems can also be fooled. The subset of electronic warfare that employs means by which these systems are defeated or deceived—to deny an adversary targeting information—falls under the category of electronic countermeasures, or ECM. There are active methods of ECM, as in the case of radar jamming, and passive methods, such as stealth technology.

Early in the miniseries, when the Raptor crewed by Boomer and Helo detected two incoming Cylon missiles, we heard:

Boomer: Jam the warheads.

Helo: I'm trying. . . . I can't find the frequency, drop a swallow.

Electronic jamming is one method of ECM in which a transmitter—the jammer—broadcasts EM radiation to interfere with enemy radar or C^3 (command, control, communications) signals. You can't shoot what you can't see. Jamming can fall into two categories: denial jamming and deception jamming. The intent of denial jamming is to overload an adversary's radar receiver, denying them the use of that device and masking the very signal that the radar was employed to detect. A jammer emits radiation at the same frequency as adversarial radar, but often at a higher power. The radar can't detect its own reflected pulses because the jammer is creating too much "noise"—as if someone stood right next to a giant searchlight and shined it into your face.

Denial jamming can be performed in several different ways—yet another element to the cat-and-mouse game of EW. Spot jamming is when a jammer broadcasts all its power to block a single frequency only. This would work if the enemy is using a transmitter of a fixed, known frequency. Rarely are radar or communications equipment tuned to a single frequency, so to counter this, some jammers spread their transmitting power over a range of frequencies simultaneously. This is called barrage jamming. The drawback to barrage jamming is that the jammer dedicates less power to individual frequencies, and can be less effective. A hybrid between spot jamming and barrage jamming, sweep jamming is a jamming technique where the full power of a jammer rapidly scans or "sweeps" over multiple frequencies. Since sweep jamming does not jam all frequencies simultaneously, its effect can be limited, or its effect can be countered. In "Exodus, Parts I and II," the resistance on New Caprica were desperate to get the "Cylon jamming frequencies." This implied that the Cylons were using either barrage or sweep jamming.

Deception techniques are those in which a repeater receives a radar pulse, modifies or delays it, and then retransmits it back to the source. This technique can be used to change the apparent position or speed of a target. An electronic device known as a blip enhancer is designed to receive a signal, amplify it, and retransmit it back to the source, essentially making it appear that the radar has detected a much larger object. What might be the purpose of making a small object appear large on radar or DRADIS? In the attack on New Caprica in "Exodus, Part II," a flight of Raptors launched two clusters of four drones with this type of technology onboard. It was a feint. The blip enhancers on the drones made it appear to the Cylons as if they were under attack by the battlestars *Galactica* and *Pegasus*, thus hiding Admiral Adama's true intent.

The decision to jam an adversary's radar has the very same perils as that associated with using active radar—whenever one combatant in a conflict radiates energy into the environment, that serves as a beacon to the enemy. Anti-radiation missiles like the HARM can home in on a radar jammer just as easily as it does on active radar. Moreover, some weapons systems of today have "home-on-jam" capability. Once the weapon senses that it is being jammed, it switches its mode of homing—instead of using its own internal radar, it uses that which the adversary has graciously provided. Missiles with home-on-jam capability are significantly harder to defeat using ECM. Therefore, there also exist passive methods to fool both search and fire-control radar. Compared to denial jamming and the forms of deceptive jamming discussed so far, some methods of mechanical jamming are comparatively low-tech.

Some rather low-tech devices can be used to reflect radar energy efficiently back to the source. This is called mechanical jamming. One radar countermeasure, called chaff, is used to create a cloud of false radar returns, usually to distract incoming radar-guided missiles. Chaff can be made of varying-length aluminum strips or metal-coated plastic or glass fiber. The length of the strips varies in order to reflect radar of different frequencies. Most warships, military aircraft, and even some ICBMs have chaff dispensers to fool incoming missiles. The warships of nineteen navies have Super Rapid Blooming Offboard Chaff (abbreviated as SRBOC or "Super-arboc") launchers aboard. This fairly simple system is comprised of a mortar that launches a huge cloud of chaff up and away from the vessel, with the intent of creating the appearance of a larger, more tempting target when incoming enemy missiles have been detected.

Decoys can be inexpensive stationary objects that readily reflect radar, or highly maneuverable autonomous flying craft. In either instance, they are designed to deceive an enemy radar operator into thinking that they are aircraft, and they can clutter radar with false targets. Decoys can be used to protect a convoy (like the Rag Tag Fleet) by creating too many radar returns, or they can make it easier for an inbound adversary to approach within weapons range. Some decoys can deploy chaff. Others even have the capability of performing deceptive jamming.

Take six mirrors. Cement them together to make a hollow cube that has the silvered sides facing inward. Saw off a corner. You have created a corner reflector, or corner cube reflector, a very simple device that has the amazing property in that it will re-radiate nearly 100 percent of electromagnetic radiation directly back toward its source. Corner reflectors placed on the moon by the Apollo astronauts are used today for lunar laser ranging—in order for scientists to determine to within a few millimeters the Earth-Moon distance and to track the Moon's orbital evolution. Corner cubes reflect radar energy efficiently, and in warfare can make an inexpensive decoy look like an attractive target. In the early stages of the attack on the Twelve Colonies, a Viper squadron from *Galactica*, with a Raptor for ECM support, approached a wing of Cylon Raiders. After processing the DRADIS signals, the Raptor's ECO, Helo, determined that the ten inbound Cylons were really only two in number—the Raiders may very well have deployed decoys with blip enhancers or corner reflectors to achieve this advantage. Had the Viper pilots known that there were only two Raiders, they would have approached more rapidly, less cautiously, and the Cylons may very well not have had time to implement their electronic attack.

There are also countermeasures to thwart missiles that employ passive homing, in particular heat-seeking missiles. When heat-seekers were first developed, an aircraft with such a missile on its tail could climb toward the Sun, a much greater source of infrared radiation than an aircraft tailpipe. Often the missile would attempt to home in on the Sun—with predictable success. Today missiles are smarter,

ELECTRONICS IN THE SPACE ENVIRONMENT

Since we speak of "the vacuum of space," we think of space as empty. It is and it isn't. There is very little matter in space, but it is full of the full spectrum of electromagnetic radiation and an entire zoo of high-energy fast-moving subatomic particles—both forms of ionizing radiation. This presents a technological hurdle in the design and operation of electronic equipment.

If a single high-energy charged particle like a cosmic ray impacts a conductor, it can ionize thousands of atoms, freeing thousands of electrons and causing electronic power surges. In the case of solid-state and digital circuitry, this can cause wildly inaccurate results and often physical damage to the equipment. The solid-state circuitry of today's spacecraft is, therefore, protected from the space environment by radiation hardening. Radiation hardening, or simply rad hardening, is a way of designing electronic components and systems that makes them resistant to damage or malfunctions caused by EM radiation and high-energy subatomic particles.

Even then, solid-state memory chips are subject to single-event upsets, or SEUs, where a high-energy charged particle can literally "flip" a bit—causing it to change from 0 to 1 or vice versa. The first 0/1 bipolar flip-flops were observed on a spacecraft in 1979, though there is evidence that they occurred on other spacecraft long before that. This phenomenon was used as a plot point in the webisodes "The Face of the Enemy," when a single-event upset in the portion of a Raptor's memory storing jump coordinates caused the Raptor to jump to an indeterminate location in space, stranding its crew and passengers. Ironically, on June 4, 2008, while the "Face of the Enemy" webisodes were being written, an SEU made the nightly news when it briefly halted the operation of the Mars *Phoenix* lander!

and deceiving heat-seekers can be a much more difficult task. Instead tactical aircraft and ships are able to launch flares. Flares burn hotter than a craft's tailpipe or a ship's smokestack, and provide tempting targets for a heat-seeking missile. In the Miniseries, when Helo told Boomer to "drop a swallow," the object that their Raptor deployed looked suspiciously like a flare that aircraft have been using since the Vietnam War to foil heat-seeking missiles.

You cannot shoot what you cannot see, and some craft, like Chief Tyrol's Blackbird, can defeat radar simply by their composition or physical geometry. The term "stealth," or low-observable (LO) technology, covers a range of techniques used to make aircraft and ships less visible, or ideally invisible, to radar and other methods of detection. Craft like the F-117 Nighthawk have very angular surfaces whose shapes efficiently reflect incoming radar energy, but simply not in the direction back to the source. Some stealth craft of today have skins made of material that reflect some incident radar radiation from their outer surfaces, and the remainder from the inner surface. The thickness of the skin is then "tuned" so that both reflections "cancel" each other out by the process of destructive interference. In some cases, as was the case with *Blackbird*, craft are simply made of materials that simply do not reflect radar (see the chapter 21 sidebar, "Finding Materials to Make the Blackbird").

Perhaps the ultimate in ECM, though, is electromagnetic pulse, or EMP. EMP is a high-intensity broadband pulse of electromagnetic radiation that can rapidly produce destructive power surges—surges that are both too rapid and too intense to be shielded by normal surge suppressors—in electronics and electrical systems like computers, radar, communication systems, electrical appliances, and automobile ignition systems. EMP damage can range from a brief functional interruption to a complete system burnout. Although lightning is one natural source of EMP, the most common source of destructive EMP is a nuclear explosion. In order to provide a sense of scale, a large nuclear detonation over the state of Kansas, at the elevation of the International Space Station, would affect electronic systems over all of the continental United States, and much of Canada and Mexico.

EMP can be generated by non-nuclear means. NNEMP (non-nuclear electromagnetic pulse) generators currently exist, and can be carried aboard bombs and missiles to create all the electronic effects of a nuclear detonation without the radioactive fallout. Recall that when Raiders attacked *Colonial One* in the miniseries, Apollo simulated the electromagnetic signature of a nuclear detonation with an electromagnetic pulse generator. Recently, it has been reported that the U.S. military has successfully managed to create EMP grenades—which would be extremely useful for covert and special force units.

Electronic Counter-Countermeasures

The term "electronic counter-countermeasures" (ECCM) refers to methods that eliminate or curtail the use of ECM by an adversary. ECCM is generally synonymous with "resistance to electronic jamming." Some modern radar, called frequency-agile radar, has the ability to continuously change the frequency on which it operates. This would defeat spot jamming, and potentially even sweep jamming.

A common technical criticism of *Battlestar Galactica* has been the use of corded phones on a technologically advanced battlestar capable of faster-than-light travel. This is actually an effective method of ECCM, since communications over a wire are far more resistant, though not entirely impervious, to electronic interception and tampering than those sent via wireless methods. Fiber-optic communication is even better. Case in point: there exist weapons in today's militaries (for example, the TOW missile and the Mk-48 torpedo of the U.S. military, or the Chinese Yu-6 torpedo) in which a weapon, once deployed, pays out a wire en route to its target—allowing jam-resistant control of the weapon until its terminal phase. Wireless intraship communications would allow the Cylons a cheap and easy source of SIGINT.

ELINT and SIGINT

Intel can also be gathered from any EM signals broadcast by an adversary like DRADIS, microwaves, or digital data communication—known as electronic intelligence (ELINT). Signals intelligence (SIGINT) refers to intelligence gathered passively through intercepted electromagnetic signals. That can be through interpersonal communications—like telephony or *Galactica*'s wireless—in which case this is known as communications intelligence (COMINT). Recall that in the episode "Tigh Me Up, Tigh Me Down," Apollo and Beehive were dispatched to intercept an inbound Cylon Raider. Apollo appeared to damage a Cylon Raider in a dogfight, and the Raider kept trying to jump away, only to jump to another nearby location still within the

Fleet. Colonel Tigh said that "this is our perfect chance to get some intel." In that case, he meant ELINT: "Order Apollo to close with the raider, but do not engage. Put a Raptor in the air. As long as that thing's flopping around out there, tell them I want to suck in every electronic signal that thing makes."

By determining the frequencies on which Raider electronics operate—the spectral signatures of its communications, search DRADIS, fire-control DRADIS, even spurious emissions from various subsystems—*Galactica* may be able to develop more effective electronic countermeasures for future engagements. In fact, recall that Starbuck and Tyrol attempted to analyze electronic emissions to determine how the Raider's FTL system worked.

Our characters were excited about the prospect of gathering ELINT from the Cylons, and it turns out that the Cylons were every bit as excited to get ELINT and SIGINT from the Colonials. It turned out that before jumping away, the Raider sent a broadcast home before being destroyed. The Raider was never severely damaged; it was collecting intel on *Galactica*. So apparently EW is not so much as a "cat-and-mouse" game as it is a "cat-and-wounded-bird" game.

CHAPTER 27

How Did the Cylons Infiltrate the Colonial Computer Infrastructure?

If you've got a certain mind-set, it was the most frightening part of the miniseries: a squadron of Viper Mark VIIs is en route to intercept an inbound Cylon armada. Just before the fighters come into range, the Cylons broadcast a signal and the onboard computer in each Viper shuts off power to the entire craft. The Vipers are dead in space, unable to fire their weapons or even maneuver. The pilots are massacred, and the Colonial Fleet has lost their first line of attack, just like that.

The Cylon secret weapon, we learn later, was a back door inserted into the Command Navigation Program (CNP)—the operating system for nearly everything that flies in the Twelve Colonies—by a consultant working for Dr. Gaius Baltar, the software's designer. A consultant who, unknown to Baltar, was Number Six, a humanoid Cylon. Having Number Six insert a secret switch into the CNP software was not a requirement for the successful Cylon attack, but it made things a lot easier. Her position as a consultant to Baltar

Cylon model Six.

Cylon model Six.

(who was himself a consultant for the defense establishment) would likely have allowed her to learn the operation codes that controlled the computer processors on craft ranging from Vipers to battlestars—information that was likely passed back to the Cylons through Cavil.

A back door is a way for an unauthorized user to gain access to a computer system. It can be as simple as deliberately inserting a special hidden user name and password into the login procedure. Usually, user names and passwords are encrypted and kept in an external database, which makes it easy for authorized users to be added to the system and expired users to be deleted. To create a back door, the programmer simply adds code somewhere along the authentication sequence saying that the user name X, password Y, is always welcome into the system. User name X never shows up on the list of authorized users, and remains active even if the entire user list is trashed and replaced. Anyone logging in with user name X is usually going to be given the same access to the system that the original programmer had, which is considerable.

Other backdoor schemes involve subverting the entire process of writing computer code. You've no doubt heard over and over that computers understand only ones and zeros. How, then, do programmers tell computers what to do? Do they write an incredibly long series of ones and zeros into the computer?

Years ago they did, but not anymore. A simple computer program looks something like this:

```
printf("Hello, World\n");
```

Even if you know nothing about computer programming, you could probably guess that this program prints the phrase "Hello, World" somewhere, either on a screen or to a printer. It's also pretty obvious that if you can understand a piece of code, a computer can't. To get that program in the example into a form the computer can use, the code has to be "translated" using another program called a compiler. The compiler takes the code and, in a few seconds, turns "Hello, World" into "01001000 01100101 01101100 01101100 01101111 00101100 10010111 01101111 10010010 01101100 01100100."

Thanks to the work of the compiler, the computer is now ready to run your program.

It is time-consuming but not very difficult to write a compiler. It is a standard assignment for undergraduate computer science majors. It is also easy to modify a compiler to compromise whatever it compiles. This is one of the most insidious ways to attack a computer system—all the code the programmers wrote is absolutely clean, yet every piece of software written on that system has a security hole, added by the malicious compiler. This method is so intelligent, logical, and destructive that it just might be what Six used against the Colonial defenses.

Galactica was being decommissioned and turned into the Colonial equivalent of the USS *Intrepid* in New York City or the USS *Midway* in San Diego. She was immune to the Cylon backdoor attack because her computers were extremely old, non-networked, and running legacy software. The non-networked part is the most important, because when computers are linked together, an attack against one computer becomes essentially a simultaneous attack against all of them. A network isn't a requirement for viruses to spread, but it makes the attack more efficient, and allows the virus to co-opt more of the computers quickly.

For this reason, and because *Galactica* computers had been networked and compromised during the First Cylon War, which led to a significant loss of life, Commander Adama vehemently insisted that the computers on *Galactica* would never be networked as long as he had anything to say about it. And because this is television, as soon as he said that we all knew that there would come a point where the computers would have to be networked.

Back doors were one way for the Cylons to gain access to Colonial computers, but not the only way. Forty years prior to the events in the miniseries, the Cylons looked more like "walking chrome toasters" than supermodels, so Cylon agents would be comparatively easy to identify. While giving a tour of *Galactica*, Aaron Doral, who we discover later is a humanoid Cylon, says: "It was all designed to operate against an enemy who could infiltrate and disrupt even the most basic computer systems."

Doral's statement implies that the Cylons are capable of accessing nearly any Colonial computer remotely—given time and reasonable proximity. The Colonials had given up trying to compete against the Cylons in computer technology, and it was generally assumed that if the Cylons wanted to hack into a computer, they were going to get in. How did the Cylons gain remote access to Colonial computers during the First Cylon War?

We can assume up front that the Colonial military is reasonably intelligent, and that a wireless network is the last thing we would find on a battlestar. A wireless network would provide the Cylons easy and unfettered access to Colonial data communications—at worst a cheap source of SIGINT (Signals Intelligence), at best easy access to Colonial computers. It turns out that the environment of attack, camouflage, deception, and countermeasure is every bit as complicated in cyberspace as on the battlefield.

It is entirely plausible that Cylons could infiltrate Colonial computer systems remotely, with surprisingly little effort, when we examine the many ways that SIGINT can be gathered today in ways that may initially seem counterintuitive. Electrons flowing through a conductor like a wire emit electromagnetic radiation. That, in fact, is a very cursory description of what an antenna does—electrons flow through the antenna and EM radiation propagates from it. Communication lines, electrical wiring, network data lines, even the electron beams of CRT monitors emit EM radiation that can be intercepted and used as a source of intelligence. More recent research has verified that the screen contents of LCD displays are similarly vulnerable.

There are other forms of EM radiation emitted by computing devices that can yield actionable intelligence. When electrons flow through the components of a circuit, heat is an unfortunate, and often undesirable, by-product. Hot components continuously radiate heat to their environment in the form of infrared radiation. If the surface of a CPU chip can be observed, it is possible to glean information about the operations being executed on that CPU from the variations in the infrared radiation being emitted by it. This method of SIGINT is known as a thermal imaging attack.

Even as far back as the early 1960s, at the height of the Cold War, spies had electronic eavesdropping devices that could tap phone conversations without physical connections to the phone's handset or wires. Devices placed near a phone line could "read" fluctuations in the electrical signals flowing through it through a process called magnetic induction. This allowed a third party to eavesdrop on the conversation. Microwave emissions due to current flowing through phone lines can be read in a similar fashion.

During the Cold War, Soviet spies even learned how to gather intelligence from one particular model of typewriter. The IBM Selectric had a motor to turn the print head. That motor drew different amounts of current depending on which key was pressed. By monitoring how much electrical current the typewriter drew, it was possible to determine what was being typed on the machine, and the document being typed could be reconstructed. If SIGINT could be gathered easily from typewriters, phone lines, and solid-state circuitry with technology that existed as far back as forty years ago, then clearly the technologically advanced Cylons could have found a way to access the information on Colonial computer systems remotely.

At the beginning of the miniseries, since *Galactica* was being decommissioned, security was beyond lax: at least one Cylon agent, Sharon "Boomer" Valerii, had been aboard *Galactica* for two years prior to the attack (and another, Doral, for at least several weeks). Gina Inviere had infiltrated Pegasus. The TV movie *The Plan* showed us that other parts of Cylon society were similarly compromised.

With the discovery of the device recovered from the DRADIS display in the miniseries, it seems that *Galactica, Pegasus*, and other ships of the Fleet were in a state similar to the U.S. Embassy in Moscow in 1987. Lax security surrounding the construction of the new embassy meant that before construction was even complete, there were an estimated $20 million worth of Soviet bugs and surveillance equipment installed in the walls and ceilings. The fiasco was an embarrassment to the U.S. government. Instead of trying to sweep the Moscow embassy building free of the surveillance equipment, knowing the procedure would likely never be 100 percent successful, the U.S. State Department demolished the embassy, improved security, then rebuilt it from scratch. Who knows how many devices, and of what nature, were implanted on the ships of the Colonial Fleet shortly before the attack? Since it was established that the devices were not bombs, it is very likely that their function was to gather intelligence.

Gleaning information from computers is one thing. Influencing those same computers is a completely different matter. However, it's not a big technological jump between an intelligence-gathering device that can "read" the data broadcast over a transmission line to one that can induce a current in the very same line.

How could they have hacked into computer systems remotely in the first war, though? Electrons flowing through a wire generate electromagnetic radiation, and EM radiation of the appropriate energy can induce a current within a wire, which is essentially how an antenna works. Given this law of physics, and given the nature of the Cylons, and even assuming that the metal hull of a battlestar might absorb some of those signals like a huge Faraday cage,* it's not unreasonable to assume that the Cylons could tap Colonial computers and networks remotely. It was established that the Cylons understood electronics very well, even during the first war, and Cylons don't eat, sleep, rest, or require recreation—so, if nothing else, they had a lot of time to devote to the problem.

Under the category of poetic justice, with Athena's help the crew of *Galactica* was able to return the favor to the Cylons by launching a virus attack of their own in the episode "Flight of the Phoenix." Now, one might think that the Cylons, who employ computer viruses as an offensive weapon, would have had the wherewithal to install antivirus software on their own military assets to prevent such an attack. Perhaps they did, but the best-laid plans of mice and Cylons often go astray.

Which leads us to another question: how much damage could a virus cause, really? Let's start with some background. In the episode "Scattered," Colonel Tigh ordered *Galactica* and the Rag Tag Fleet to perform an emergency FTL jump, believing a Cylon attack was impending. Unfortunately, due to a procedural error, *Galactica*

---

* A Faraday cage is a box covered on all sides by grounded fine-mesh metallic screens that absorb radio waves. Anything inside the box is effectively cut off from all electromagnetic interference in that part of the spectrum. Cell phones can't get a signal, and radios play nothing but static, if that. Build one—it's freaky.

jumped to one set of coordinates, the rest of the Fleet to another. The crew of *Galactica* had to jump the ship back to their previous coordinates in order to obtain new celestial fixes so they could, in turn, determine the actual jump coordinates of the Fleet. In order to perform these calculations more rapidly, knowing that the Cylons likely would be laying in wait, they networked the computers from several of *Galactica*'s subsystems to create a parallel computer. Instead of working on the mathematics one step at a time, as normal computers do, a parallel computer can work on several steps simultaneously, and, Gaeta hoped, come up with a solution in a fraction of the time.

Galactica had to fend off both a Cylon military attack and an electronic attack while the computers did their number-crunching. Although Gaeta was smart enough to install software firewalls between the computers, the Cylons were too wily for him. Almost instantly, the Cylons started an intrusion attack against *Galactica*'s network and, as Colonel Tigh said, "it's now a race" to see if Gaeta's firewalls could fend off the Cylon attack long enough to complete the FTL calculations. Since there was always the assumption that the Cylons would get in eventually, it wasn't a matter of keeping them out, it was a matter of how long they could be kept out. As you can see if you watch the episode carefully, the Cylons get their virus in just under the wire, and that virus infests the networked computers, to return later in "Flight of the Phoenix."

A computer virus is a piece of software that, in its purest form, simply replicates itself across a computer network, much in the same way a biological virus spreads itself through a population. A biological virus inserts its genetic material into the nucleus of a cell, "reprogramming" it to create copies of the virus instead. A software virus inserts its code into the CPU of a computer, programming it to create copies of the virus in other programs. There are even computer viruses that work like biological retroviruses: they rewrite their own code with each infection, making them extremely difficult to detect. Like a biological virus, a computer virus can be fast-moving and virulent, or slow and relatively benign.

In real life, almost all computer viruses carry some sort of payload—another hunk of code that is activated to behave maliciously once the virus has gotten into your computer. One of the most common payloads is an engine that turns an infected computer into a spam machine. Those e-mails you receive offering "v1agra" and "naked teens who take their cl0ths off 4 U" might well be coming from your own computer.

If only Gaeta were so lucky as to have downloaded a Cylon spam machine. Shortly after infection, the Cylon viruses in *Galactica*'s computers started to flex their collective muscles. Dr. Baltar inferred that the viruses had likely been running in parallel on every computer on *Galactica*. So while it was the network that allowed the viruses to propagate between computers, the viruses can both run and evolve independently. The viruses began to shut down systems across the ship, probably in preparation for their ultimate action: opening all the airlocks and venting the ship's atmosphere into space. Gaeta immediately got to work and managed to eliminate the virus before that happened, but not before the virus planted a logic bomb in the code.

A logic bomb is a piece of software inserted into a computer system that activates itself at a later date, or when other specific conditions are set. A logic bomb can be as simple as a switch that disables a software package if it's not paid for, or a job-security monitor that deletes every hard drive in the company if the programmer who wrote the logic bomb is ever fired.

The U.S. Department of Homeland Security is, frankly, petrified of logic bombs. They foresee a very slow foreign terrorist cyberattack in which logic bombs are quietly planted by viruses or human agents into large-scale SCADA* systems across the United States. At a given signal, all those logic bombs would be triggered at once, leading to Zero Minute: computer crashes at airport control centers, multiple citywide traffic jams, regional power blackouts, and nuclear reactor accidents all happening across the United States at exactly the same time. A well-coordinated attack, like the one Six managed to launch against the Colonial defenses, could damage or destroy a great deal of the infrastructure of the nation.

---

*Supervisory Control and Data Acquisition: the collective name for the computer systems that control traffic lights, sewage systems, factories, air traffic control towers, nuclear reactors, and so on—all the industries and facilities that keep our lives going.

CHAPTER 28

So Where Are They?

In the series finale, the three-hour "Daybreak," we realize that the Galaxy should have at least two other samples of intelligence aside from our own: the sentient Centurion civilization, set free by the Cylons to go their own way, and the Being(s?) that created the apparitions we call Head Six and Head Baltar. Could that be it? In a universe that seems expressly built to allow life, in a Galaxy of hundreds of billions of stars with maybe a trillion solid worlds, could there really be only three intelligent civilizations?

If the answer is no, then where are the rest of the intelligences in the galaxy?

If the answer is yes, then what does this lack of intelligence mean?

Nick Bostrom, a philosophy professor and director of the Future of Humanity Institute at Oxford University, thinks he might have the answer. And it might be that the

Lee "Apollo" Adama and Kara "Starbuck" Thrace.

Lee "Apollo" Adama and Kara "Starbuck" Thrace.

apocalyptic visions of *Twelve Monkeys, Soylent Green, Deep Impact, The Day after Tomorrow, Planet of the Apes*, or *Battlestar Galactica* itself could be everyday occurrences throughout the Milky Way. The discovery of extraterrestrial life, he claims, is probably the worst news our civilization could receive. The more complex that life is, the worse things are. If Bostrom had been part of *Galactica*'s Fleet, he might argue that the discovery of Kobol, New Caprica, the Algae Planet, Dead Earth, and Regular Earth all add up to doom for civilization.

His argument goes something like this: Ours is not the first star system created in the galaxy; there are many millions of planetary systems older than our own. If those earlier planetary systems followed the same evolutionary path we did, they would have developed intelligence long ago. But there's no evidence that they did. Why not? Bostrom argues that there are some highly improbable steps, which he calls barriers or filters, in each stage of the development of intelligent life. There's a barrier that must be overcome to create a planet. There are barriers that must be overcome to seed that planet with the appropriate chemicals for life. There are barriers against life forming, and multiple barriers on the way toward developing intelligence. There are even barriers against developing spacecraft that can colonize the galaxy. Bostrom unifies these barriers into something he calls the Great Filter, a set of circumstances that "prevents the rise of intelligent, self-aware, technologically advanced, space-colonizing civilizations."

Bostrom suggests that the best possible situation is if the Great Filter, the thing that prevents us from exploring the galaxy, is behind us. If that is the case, then we've already overcome whatever roadblocks to intelligence the universe has placed in our path. Is it hard to get life started? We've done it already. Hard to get life out of the oceans? We've done it already. Really hard to develop intelligence? BT, DT. Having passed through an unlikely combination of hurdles, we may find ourselves lonely in the universe, but at least we're here, with a limitless future before us.

But what if the opposite is true? What if life is as easy to start as a fistfight in a bar? What if planets all over the galaxy teem with algae and trees, with amphibians and reptiles, as seems likely given *Galactica*'s depiction of the Milky Way? In that case, each new life form, each level of biological complexity, would be another toll of doom for our species.

There must be some reason why we haven't heard from anyone else in the galaxy, Bostrum argues. If we were to discover a bunch of algae planets, we'd know that the Great Filter most likely lies after the creation of algae. But how far after? If we somehow discover planets with the equivalent of a terrestrial biosphere of land-based animals, we'd know that the Great Filter most likely lies after the creation of land-based animals. But again, how far after? And if the Milky Way is festooned with intelligence but not with interstellar travel—civilizations at roughly the level we are right now—there must be something that prevents intelligent life from exploring the galaxy.

If that's the case, Bostrom argues, then the Great Filter lies ahead of us. And this filter is not some paltry technological barrier, Bostrum claims. If it turns out that FTL drives or hyperspace engines are flat-out impossible (as it looks now), any developed civilization can send out colony ships using ordinary rocket propulsion. They'll take centuries to reach the nearest star, but they'll get there.

Suppose it takes a civilization takes ten thousand years to develop its first slow interstellar ship, a ship much like the ship the Final Five used to escape Dead Earth. It sends this ship out on a two-hundred-year mission to the nearest star, where the inhabitants establish a colony. If the home planet and each colony then send out their own slow ships at two-hundred-year intervals, most of the galaxy could be colonized in about a million years. If it is so easy to develop intelligence and slow interstellar travel, such a civilization should have formed long before ours. And since it is possible for even the slowest of them to have colonized the galaxy by now, then where are they?

Bostrom is so optimistic about our ability to work around any technological barrier to space exploration that he claims that a future Great Filter must be an existential one. Whatever it is that stops a civilization from exploring the stars, he says, must stop the civilization entirely. (Perhaps the mantra of *Battlestar Galactica*, "All this has happened before and will happen again," is just another way of stating Bostrom's Great Filter.) It may be that humanoid intelligences invariably decide to play with fate and create the objects of their own destruction. Perhaps our civilization, or any civilization, is doomed to cycle through rises and falls at 150,000-year intervals. Maybe intelligence, in the long run, doesn't have survival value.

Of course, there are some holes in Bostrom's argument. Maybe we just happen to be the first ones in the galaxy to evolve intelligence. Someone had to be. Maybe Bostrom's great technological optimism keeps everyone at home in the same way some people delay buying a new computer for years because "the new models will be better, and cheaper," perhaps civilizations are loathe to invest in slow colony ships because they constantly hope for the next breakthrough in relativistic propulsion, which never comes.

But if these technological stay-at-homes exist, they still should have discovered radio. We should be able to point our radiotelescopes at star after star and hear other civilizations chattering throughout the cosmos. Yet we don't. Our fifty-year SETI program, in which we listen for extraterrestrial radio signals, has so far turned up nothing of value in all its various incarnations. A recent development that tangentially deals with the *Galactica* franchise might give some idea why.

In 2009, between the last episode of *Battlestar Galactica* and the first episode of *Caprica*, the United States converted from analog television broadcasting to digital television broadcasting. This change was bothersome for some people with older TV sets, went unnoticed by a great many more, and was a boon to the makers of digital television broadcasting and decoding equipment. But it also drastically changed Earth's radio signature with the flip of a switch.

In the analog days, the most raucous chunks of radio real estate were the television frequencies. An analog TV signal contains a series of regular pulses mixed with an easy-to-understand modulated signal. Any aliens who could pick up our TV signals would know that there was *something* intelligent there, even if they weren't able to decode the whole thing at first. But analog transmission is wasteful. You can fit much more information into a digital signal, with one drawback: a well-compressed digital signal is virtually indistinguishable from static, or noise. Unless you know the rules to decode that specific digital signal (and the number of rules is literally infinite), you might not even know there is a signal there! If there are friendly next-door extraterrestrials, and if they have been broadcasting digitally all along, we *might not ever detect them*. The universe could be swarming with intelligence all around us, and if they're broadcasting digitally, we might not know it unless they lowered themselves to speak with us.

So what does this say for our place in the cosmos? Large-scale TV broadcasting really started on Earth in the 1950s. There's now a bubble around Earth, nearly sixty light-years in radius, containing the carrier signals of all our TV shows from *Kukla, Fran and Ollie* to the *Battlestar Galactica* finale. With European television in the process of going all-digital, and with Asia still working out the details to go all-digital sometime in the 2020s, it's possible that by, say, 2050 Earth will cease to spew any analog broadcast TV into the cosmos.

So maybe Bostrom's filter is a cognitive one, based on the switch from analog to digital. Maybe there are advanced civilizations hiding on the other side of the digital filter, waiting for us to stop our 100 years of analog yammering. They could be looking for that special lack of a signal that tells them we're worth talking to.

AFTERWORD

Richard Hatch as Tom Zarek.

By now you've probably already concluded that science did play an important role for Ron Moore and the writers of *Battlestar Galactica*. Unlike so many other science fiction movies and programs, *BSG* used science not as a veneer, but as a key thematic component for driving many of the character stories—whether it was Chief Tyrol surviving the death-defying journey out the airlock, or the way craft were shown maneuvering through space, the writers artfully used physics to anchor and accentuate the dramatic possibilities and create a realistic context for the show. *BSG* didn't shy away from science and take the lazy man's way out, but, rather, used it to build context and to heighten and emphasize the dramatic themes, which is the art of science fiction in my opinion.

We now live in a world of fusion where the past, present, and future can all blend together, creating even more exciting and powerfully moving dramatic options, and *BSG* certainly is at the forefront of building a bridge between contemporary drama and the unpredictable world of future possibilities in science fiction. Although the science and physics of the future can be fascinating to many of us and a vitally important aspect to any well-written science fiction program, those of us who have loved the genre for decades will tell you that the best has always been about exploring both theoretical technical and scientific possibilities and the mysteries of the human heart—as any good character-driven drama does.

With courage, healthy audacity and, thank gods, the support of the Syfy Channel, *BSG* was successfully able to blend and integrate the very best of science fiction with more conventional themes; moving from the black-and-white dramatic scenarios of the past to the more honest, ambiguous, and morally conflicted themes to which we, in this more illuminated world of today, can relate. *Battlestar* reminded us not only of our deepest flaws and fallibilities, but of our humanity as well. It is rare when a science fiction series not only can entertain, but inspire, expand our viewpoint of the world, and be socially and politically relevant to the times in which we live. The series not only mirrored much of what has been going on in the world over the past several years, but amplified and focused penetrating insight into how we human beings deal with catastrophic life-and-death events.

The show portrayed complex and yet simple human beings dealing with the sociological, political, and physiological challenges of surviving catastrophic change, which we in this world of today are also facing. When familiar dramatic themes are seen in the light of future possibilities, new windows of perception are opened up. Given that the historical topography of the series was only one step removed from our present-day reality, the writers were able to delve into the show's much darker apocalyptic scenarios with a great deal more honesty and integrity and with less fear of alienating their audience than they would were the show in a modern-day setting. The show found a populist appeal and fans who had never watched, or been interested in, such a "geeky" genre were converted and assimilated in droves as *Battlestar* moved at an ever-accelerating pace to its final chapter and closure.

As for me, I will never see *BSG* as having ended or in its final resting place, only suspended for a short while until the Powers That Be recognize how many fans across genres love and support this amazing story. *Battlestar* has not only managed to redefine science fiction in the eyes of viewers and critics alike, but pushed the limit of dramatic possibilities. May *Battlestar Galactica* live forever in our hearts and in our imagination!

So Say We All!!

Richard Hatch

NOTES

Introduction: Moore's Law

1. From the lyrics of "Mystery Science Theater 3000 Love Theme" by Joel Hodgson and J. Elvis Weinstein, used by permission.

1. Are You Alive?

1. One interesting side note: As we examine more and more primitive life forms and the more primitive areas of current genomes, we find that older organisms and older genes rely on those amino acids that can be most easily created in Miller-Urey–type experiments. This hints, but does not prove, that the genetic code for early organisms was biased toward a smaller number of amino acids—the amino acids that can be found in a world like Stanley Miller's ocean flask.

5. How Can Cylons Download Their Memories?

1. Issue 16.07: Infoporn: Tap into the 12-Million-Teraflop Handheld Megacomputer: http://www.wired.com/special_multimedia/2008/st_infoporn_1607.

9. Energy Matters

1. Called BIPM (the Bureau International des Poids et Mesures) by the French.

11. Special Relativity

1. Einstein, Albert. "On the Electrodynamics of Moving Bodies." *Annalen der Physik* 17: 891–921 (1905). In the original German, if you're feeling ambitious. English translations are readily available online, at http://www.fourmilab.ch/etexts/einstein/specrel/www/, for example. The math isn't *too* difficult, if you're up on your algebra.

12. General Relativity and Real Gravity

1. In SI units, the value is 6.674×10^{-11} m^3/kgs^2.

14. The Effects of Nuclear Weapons

1. Some elements of this list come from the Web page "Types of Nuclear Weapons," http://nuclearweaponarchive.org/Nwfaq/Nfaq1.html, by Carey Sublette.

15. Our Galaxy

1. There is another, larger, catalogue used by astronomers called the New General Catalogue (NGC). The Andromeda Galaxy, M31, is also known as NGC 224. The Colonials obviously have something similar. In the episode "A Measure of Salvation" Lee refers to the "New Colonial Database."
2. Some recent estimates range as high as a trillion.

16. A Star Is Born

1. The actual formula is $L/Ls = (M/Ms)3.5$, in which L and M are the luminosity and mass of a star, and Ls and Ms are the luminosity and mass of our Sun. The great thing about this formula is that you don't need to know the actual values of the mass of the Sun or the other star. Simply state that the mass and luminosity of the Sun are both equal to 1, and the mass or luminosity of the other star is equal to some multiple of solar values: 10 times the Sun, 1/5 the Sun, etc.

18. Black Holes

1. These quotes and other great observations about black holes can be found in "Monster of the Milky Way," an episode of the PBS series *NOVA*.

20. Water

1. We would find life claustrophobically limited if all of our chemical processes only took place when the ambient temperature was between 65 and 73 degrees Fahrenheit.
2. Most likely on the tyllium mining ships.

CREDITS

All photographs © NBC Universal Photo with the exception of those listed below:

Page 47, bottom: Image courtesy of CERN
Pages 109, 143, 240, 242: Image courtesy of NASA/JPL/Caltech
Page 110: Figure courtesy of NASA
Page 146: Image courtesy of NASA/StSci
Page 295, bottom: Peggy Sue Davis

ABOUT THE AUTHORS

Patrick Di Justo

Pretending to represent Gemenon at the United Nations panel on *Battlestar Galactica*.

When not writing about *Battlestar Galactica*, Patrick Di Justo is a contributing editor for *Wired* magazine and has written for *Popular Science, Scientific American, New York* magazine, and *New York Times Circuits*. He has been an astrophysics lecturer at New York's Hayden Planetarium, a robot programmer for the Federal Reserve Bank, the Son of Sam's paper boy, and a standup comedian. He designed and built experiments that flew on Space Shuttle Flight STS-107.

Dr. Kevin R. Grazier

On the CIC set in Vancouver, Canada.

Dr. Kevin Grazier served as the Science Advisor for all four seasons of *Battlestar Galactica* and is currently in that role on the SyFy Channel series *Eureka* and the NBC animated series *The Zula Patrol*. He has also consulted on numerous other books, movies, and television series, including several currently in various stages of production.

Dr. Grazier is a Research Scientist at NASA's Jet Propulsion Laboratory (JPL) in Pasadena, California, currently working on both the Cassini-Huygens mission to Saturn and Titan and the Constellation program. At JPL he has written mission planning and analysis software that won both JPL- and NASA-wide awards. He worked previously at the RAND Corporation, processing data from the Viking missions in support of the Mars Observer mission.

He has BS degrees in computer science and geology from Purdue University and another in physics from Oakland University. His doctoral research was in planetary physics at UCLA, where he performed long-term large-scale computer simulations of early solar system evolution dynamics, and chaos—research that he continues to this day. In what passes for his spare time, Dr. Grazier teaches at both UCLA and Santa Monica College—classes in basic astronomy, planetary science, cosmology, the science of science fiction, and the search for extraterrestrial life.

Dr. Grazier was the recipient of Oakland University's 2009 Odyssey Award for the alumnus whose life exemplifies Oakland University's motto: "Seek virtue and knowledge."

INDEX

Note: Page numbers in *italics* refer to photos and illustrations.